DATE DUE		
MAR 0 6 1995		
ILL 6264195		
WITHDRAWN		

TRY TO IMAGINE a spaceship that could pass right through the Earth without even noticing it was there; a spaceship that could cross the vastness of space at the speed of light, and then penetrate into the very heart of subatomic matter to seek out its fundamental structure. Imagine, then, a particle that is almost nothing that can tell you almost everything about the structure of matter and the evolution of the Universe. Impossible?

In fact, all of these descriptions can be applied to the neutrino, a subatomic particle that is so elusive it is almost undetectable. *Spaceship neutrino* charts the history of the neutrino, from its beginnings in the 1930s, when it was postulated as a way of explaining an otherwise intractable problem in physics, to its crucial role in modern theories of the Universe.

Christine Sutton is well known for her popular science writing. In this book she describes how the detection and measurement of neutrino properties have tested technology to its limits, requiring huge detectors, often located deep in mines, under mountains or even under the sea. As part of the story she explains without the use of mathematics how our understanding of the structure of matter and the forces that hold it together have come from work with neutrinos, and how these apparently insignificant particles hold the key to our understanding of the beginning and the end of the Universe.

This fascinating, well written and highly illustrated book will be enjoyed by anyone with an interest in modern physics or astronomy, from school level right through to the professional scientist.

SPACESHIP NEUTRINO

SPACESHIP NEUTRINO

CHRISTINE SUTTON

Department of Physics
University of Oxford

CAMBRIDGE
UNIVERSITY PRESS

Published by the Press Syndicate of the University of Cambridge
The Pitt Building, Trumpington Street, Cambridge CB2 1RP
40 West 20th Street, New York, NY 10011-4211, USA
10 Stamford Road, Oakleigh, Melbourne 3166, Australia

© Cambridge University Press 1992

First published 1992

Printed in Great Britain at the University Press, Cambridge

A catalogue record of this book is available from the British Library

Library of Congress cataloguing in publication data

Sutton, Christine.
Spaceship Neutrino/Christine Sutton.
   p.   cm.
Includes bibliographical references (p.) and index.
ISBN 0-521-36404-3 (hc). – ISBN 0-521-36703-4 (pb)
1. Neutrinos. 2. Matter – Structure. 3. Cosmology.
4. Astrophysics. I. Title.
QC793.5.N42S88 1992
539.7′215 – dc20   92-4215   CIP

ISBN 0 521 36404 3 hardback
ISBN 0 521 36703 4 paperback

CONTENTS

FOREWORD

I have done a terrible thing, I have postulated a particle that cannot be detected.

W. Pauli

Oh Pauli, Fermi guide us,
  Banish our illusions
  Elevate our hunches
To sensible conclusions.

When the neutrino was postulated by Wolfgang Pauli and made into a beautiful theoretical construct by Enrico Fermi it was not imagined that it would reveal such a rich harvest in our understanding of the Universe. What was first introduced as an 'apology' for the apparent nonconservation of energy/momentum in beta-decay has, by virtue of the very weakness of the interaction which made its direct detection such a challenge, become a unique probe of stellar interiors!

The difficulty of transforming the 'poltergeist' to a reality was outlined by H.A. Bethe and R.E. Peierls who, using Fermi's theory, calculated the miniscule interaction distance to be measured in light years of lead. With the neutrino sources and detectors then available (1934), the detection of neutrinos was pronounced to be impossible.

Two developments changed the situation in a dramatic way: the discovery of the fission process and the development of giant liquid scintillation detectors. Given the consequent increase in detector sensitivity and an enormously enhanced neutrino flux, it was possible for us to obtain a predicted neutrino interaction rate of several per day and to make a definitive series of tests of its existence (Clyde Cowan and Fred Reines).

At this point (1956) it was recognised that the concurrent discovery of parity nonconservation by Lee, Yang and Wu required the predicted interaction cross-section to be increased by a factor of two, a result found when the experimental accuracy was improved. It is interesting to conjecture how the community would have reacted to the nonconservation of parity in weak interactions if the free neutrino had not been detected!

Given the existence of the neutrino, it was reasonable to ask what properties it possessed, such as rest mass, magnetic moment, equality of v and $\bar{v}$, identity of v_e and v_μ, stability, etc., etc. and indeed such questions have engaged the attention of particle and astrophysicists to this very day.

As described in this book the flowering of neutrino physics with sources ranging from reactors to accelerators to cosmic rays, the Sun and a supernova, has made it an exciting field of investigation. Further, the reader is well served by the perspective given of the relationship between the neutrino and other branches of particle physics.   *F. Reines*

PREFACE

Neutrinos are for me the most fascinating aspect of a fascinating subject. They are the most amazing, paradoxical, intriguing – choose whatever adjective you like – of all elementary particles. One of the most paradoxical aspects of neutrinos is that although they consist of very little and are very difficult to detect, they have taught physicists a great deal. Indeed, it is probably possible to cover all aspects of particle physics from the starting point of neutrinos. That would, however, take several volumes (and more patience than I could ever expect from my family), so instead I have tried to concentrate on the various aspects that I find most interesting. In particular, there is more emphasis on experiments to study neutrinos than on theoretical ideas and speculations about neutrinos. This is because I have tried to emphasise what we know about neutrinos and how it is that we know it. It means that I am more or less guaranteed to upset virtually everyone who has ever worked with neutrinos, either by not giving enough detail about their theory or experiment, or more likely by omitting it completely!

I must also confess that my enthusiasm is in a sense once-removed, in that I have never worked on an experiment with neutrinos. However, I have endeavoured to overcome my shortcomings by enlisting help from many experts around the world. In particular I would like to thank the following for their invaluable help and encouragement which enabled me to persevere through a project that took longer than I ever foresaw. In the spirit of modern research papers in particle physics this is a long list, in alphabetical order:

Finn Aeserud (Niels Bohr Archive)
John Bahcall (Institute for Advanced Study, Princeton)
Milla Baldo-Ceolin (University of Padova)

Ralph Becker-Szendy (University of Hawaii)
Norman Booth (Oxford University)
David Caldwell (University of California, Santa Barbara)
Donald Cundy (CERN)
George Ewan (Queen's University, Ontario)
David Faust (SLAC)
Gordon Fraser (CERN)
Georgio Giacomelli (University of Bologna)
Maurice Goldhaber (Brookhaven National Laboratory)
Marjorie Graham (American Institute for Physics)
Petra Harms (DESY)
Nigel Henbest (Hencoup Enterprises)
E. Holzchuh (University of Zürich)
Cecilia Jarlskog (University of Stockholm)
Henry Kendall (MIT)
Till Kirsten (Max-Planck Institut für Kernphysik, Heidelberg)
Maso-Toshi Koshiba (Tokai University)
Walter Kundig (University of Zürich)
John Learned (University of Hawaii)
Leon Lederman (University of Chicago)
Michael Moe (University of California, Irvine)
John Mulvey (Oxford University)
Gerald Myatt (Oxford University)
Martin Perl (SLAC)
Roger Phillips (Rutherford Appleton Laboratory)
Bruno Pontecorvo (JINR Dubna)
Roswitha Rahmy (CERN)
Frederick Reines (University of California, Irvine)
Fred Rick (Los Alamos National Laboratory)
Michael Riordan (SLAC)
Subir Sarkar (Oxford University)
Janet Sillas (Brookhaven National Laboratory)
Brian Southworth (CERN)
Bernhard Spaan (DESY)
Christian Spiering (Institut für Hochenergiephysik, Zeuthen)
David Wark (Oxford University)
Richard West (ESO)
Stan Woosley (University of California, Santa Cruz)

I would also like to acknowledge the valuable resources of the libraries at
CERN, the Rutherford Appleton Laboratory, Oxford University Nuclear
Physics Laboratory, and the Bodlean Library, Oxford; also assistance from
the Photographic Group in the Physics Department at Oxford. Lastly, I am
indebted to Dino Goulianos from Rockefeller University for unwittingly
giving me the idea for the title of the book, and, by no means least, to Terry
and Steve for their tremendous support, allowing me to spend long periods
hidden in 'Mum's' study.

1

Introduction

Neutrino physics is largely an art of learning a great deal by observing nothing.[1]

Haim Harari, 1988.

ON 23 FEBRUARY, 1987, a hoard of billions upon billions of extragalactic messengers swept through the Earth. Only a few stopped at our rocky planet; the vast majority continued onwards, unperturbed on their journey across space.

The messengers were *neutrinos*, subatomic particles with few measurable properties, barely more than pieces of nothing. Yet neutrinos are one of the dominant forms of matter in the Universe, outnumbering the more familiar particles of atomic nuclei by a billion to one. Neutrinos produced in the Big Bang, the explosive event from which our Universe appears to have grown, permeate every cubic centimetre of space. Moreover, the Sun bathes our planet with neutrinos just as it bathes it with sunlight. Hold out your hand as if to catch them and, whether it is night or day, each second ten thousand billion solar neutrinos will pass through it.

Like the Sun, all stars emit neutrinos, but the extragalactic neutrinos that swarmed through the Earth early in 1987 marked a special event. Some 170 million years ago, in a nearby galaxy called the Large Magellanic Cloud, a star nearly 20 times the size of the Sun died in a catastrophic explosion, flinging subatomic debris out into space. On 23 February, 1987, the first signals from the explosion reached Earth – myriads of neutrinos bearing witness to the death of the distant star. A few hours later, astronomers in the Southern Hemisphere saw the first light from the explosion, heralding the appearance of SN1987A, the first supernova to be visible to the naked eye for 400 years.

The detection of only a few of the neutrinos from the supernova marked the beginning of a new branch of observational astronomy. It also reinforced the growing relationship between particle physicists, who study the small-scale, barest essentials of matter, and astrophysicists and cosmologists who study the workings of the Universe at large. Neutrinos form a substantial, albeit almost invisible, bridge between one of the oldest sciences, astronomy, and one of the newest, particle physics. Whereas astronomy has its roots in antiquity, particle physics emerged as a full branch of science only in the early 1950s, the progeny of research into the structure of the atom. My intention in this book is to show how neutrinos came to play a leading role in particle physics, eventually forging links to astronomy and cosmology in a variety of fascinating ways.

Anyone who is familiar with the basic ideas of particle physics can now turn quickly to the beginning of Chapter 2. Others may welcome the brief description that follows here of the place of neutrinos in particle physics. I have attempted to introduce the necessary concepts at appropriate places in later chapters, nevertheless some of you may find it helpful to be given some preliminary guidance as you embark on your journey into the realm of neutrinos.

During the first decade of the twentieth century, physicists discovered that atoms consist of a tiny, central positively charged nucleus, surrounded by a cloud of negatively charged electrons. They went on to learn about the structure of the nucleus and found that it is built from positive particles, which became called protons, and neutral particles, which were dubbed neutrons. The charge on the protons exactly balances that on the electrons, so that a complete atom, in which the number of electrons equals the number of protons, is electrically neutral. But in most other respects, protons and neutrons differ dramatically from electrons. Most noticeably, protons and neutrons are much heavier, being around 2000 times as massive as electrons.

In the world about us, protons, neutrons and electrons combine together to form more than 90 different kinds of atom, each unique to a specific chemical element. From these foundations stems Earth's great diversity, as atoms of different elements react together to make compounds – water, for example, consists of hydrogen and oxygen atoms locked together in precise proportions, while vastly more complex structures of hundreds of atoms form proteins, the building blocks of life itself.

Most of the matter on Earth is completely stable, but some atoms are intrinsically unstable – we say they are radioactive, because they radiate energy as they change from one form to another. Radioactivity, discovered late in the nineteenth century before the structure of the atom had been unravelled, is a natural phenomenon, occurring, for example, to varying degrees in the rocks beneath our feet.

In all types of radioactivity the underlying process is the same: an atomic nucleus spontaneously changes to a less energetic state, emitting the excess energy as it does so. This energy emerges as radiation – sometimes as gamma-rays, a highly energetic form of light; sometimes in the form of alpha-particles, which are ultra-stable configurations of two protons and two neutrons; and sometimes as pairs of electrons and neutrinos. The neutrinos

Wolfgang Pauli (1900–58) –
'father' of the neutrino. (AIP Niels
Bohr Library, Goudsmit
Collection.)

produced in this way contribute billions to the swarms that traverse our environment.

It was studies of radioactivity that led to the first suggestion of the existence of the neutrino in the early 1930s. Investigations of the radioactive transmutations that emit electrons suggested that some of the energy was missing. Wolfgang Pauli, a brilliant Austrian theorist, proposed that an electrically neutral, lightweight, very feebly interacting particle must take away some of the energy released. Two decades later, researchers in the US finally succeeded in observing these particles through their extremely rare interactions with matter. And so began experiments with neutrinos, which have led not only to observations of neutrinos from the Sun and the supernova SN1987A, but also to investigations of subatomic particles and their forces. One of the most paradoxical features of these extraordinary particles is that although they are so difficult to detect, passing readily through the Earth without interacting, they have revealed vital information about almost all aspects of subatomic particle physics.

That neutrinos can be detected at all is due to the fact that they *can* interact occasionally with other subatomic particles, through the agency of one of nature's fundamental forces, the *weak force*. This is the force that underlies the radioactive transitions that lead to the emission of electrons and neutrinos. In atoms, the weak force is at least 1000 times more feeble than the electromagnetic force, which acts between charged particles and keeps the atom bound together. And it is a 100 000 times weaker than the *strong force*, which binds protons and neutrons together in the nucleus.

The weakness of the weak force means that neutrinos can travel cosmic distances without interacting with other matter. But fortunately for physicists, the whole process is rather like life itself. In developed countries most of us have a good chance of surviving at least our 'three score years and ten'. However, some people live for only a few years, while others can last a century a more. So it is with neutrinos travelling through a large detector. While the likelihood is that a neutrino will travel through the detector unaffected, there is a small but real probability that it will interact through the weak force with one of the subatomic particles in the material there. By studying these weak interactions of neutrinos, particle physicists have learned not only about the neutrinos themselves, but also about the particles with which they interact and the force through which they interact.

In the 1970s, experiments with neutrinos, in addition to complementary experiments with electrons, revealed a new level of subatomic matter. It turns out that protons and neutrons are not themselves fundamental, but consist of smaller particles, the *quarks*. The quarks are bound by the strong force within the protons and neutrons, and while single quarks cannot be knocked out of the larger particles, it is possible to 'see' them inside the protons and neutrons by sending in neutrinos or electrons. Rather as radar reflections reveal aeroplanes or ships hidden in dense fog, so the scattering of neutrinos or electrons reveals the tiny quarks buried deep within protons and neutrons.

So far there is no experimental evidence for structure within quarks themselves, and it is possible that they are truly fundamental building blocks of matter. Similarly, electrons and neutrinos appear to be simple, structure-less particles, differing from the quarks only in that they do not feel the strong force. Over the past two decades, the idea has grown that there are two distinct 'families' of fundamental particles: those that feel the strong force – that is, the quarks – and those that do not, which are called *leptons*.

The existence of matter here on Earth depends on two kinds of quark, required to build protons and neutrons, and two kinds of lepton – the electron and its related neutrino, which is emitted in radioactive decays. But as long ago as the 1930s studies of cosmic-rays began to show that Nature requires more particles than the four needed here on Earth. Cosmic-rays are energetic particles, including neutrinos and gamma-rays, which stream down through the atmosphere. While many of the neutrinos originate with the Sun, or cataclysmic events such as SN1987A, other neutrinos, and many charged particles, are produced in nuclear reactions that occur when very energetic particles from outer space (mainly protons) collide with atoms in the upper atmosphere.

Most of the charged particles produced in the atmosphere are short-lived and decay, rather as radioactive nuclei do. After only several billionths of a second, they transmute to more stable particles, eventually yielding electrons, protons and neutrinos. These short-lived particles can also be produced in controlled conditions here on Earth, in experiments at laboratories that house particle accelerators. These machines take beams of particles, such as protons or electrons, and feed them energy from electric fields, accelerating them to velocities close to the speed of light. When the

Experiments with high-energy neutrino beams at CERN, the European centre for research in particle physics, have made several important discoveries about the fundamental particles and forces. The laboratory lies on the northern outskirts of Geneva, the main site occupying the centre foreground on this aerial view which is looking south towards Mount Blanc. (CERN.)

energetic particle beams strike matter, they produce new short-lived particles, just as in the cosmic-ray collisions in the atmosphere.

Studies of these short-lived particles have shown that there are more kinds of quark than are needed to build protons and neutrons, and more types of lepton over and above the electron and its related neutrino. At present all the evidence points to there being six types of quark, and six types of lepton, of which three are charged (and include the electron) and three are neutral. All three types of neutral lepton are called neutrinos.

The interactions of all these particles through the strong, weak and electromagnetic forces are described within a theoretical framework known as the *Standard Model*. During the 1970s, neutrino experiments at particle accelerators played an important part in establishing this model, which describes very well the interactions of the fundamental quarks and leptons. Yet good as it is, the Standard Model has many shortcomings. For example, it does not include the fourth fundamental force, gravity, nor does it explain why the masses of quarks and leptons are what they are. Neutrinos of all three types have masses far smaller than their charged lepton counterparts, but are their masses indeed zero? Experiments so far cannot say for sure, and the Standard Model cannot tell us.

The role of neutrinos as tools of particle physics is not yet over, for they will almost certainly help to guide theorists in improving the Standard Model, just as they helped to reveal the nature of quarks and the weak force. But with the detection of neutrinos from SN1987A, these remarkable particles are also becoming tools for astronomers. Neutrino 'telescopes' may, over the next decade, reveal the source of ultra-high energy cosmic-rays, long a puzzle to astrophysicists.

Neutrinos, which are about as close as something can come to being nothing, can not only tell us about much of particle physics but can also provide a unique window on astrophysics and cosmology. This book will take you on a tour of the ways in which physicists have learned so much through studying neutrinos and of the struggles involved. Some of it, like neutrino experiments themselves, will not be easy, but I hope that like the experimenters, you will persevere and come to regard neutrinos with the same fascination that I, and many others, have for these elusive particles. The journey is going to take you to the heart of the atomic nucleus, deep into outer space, and back in time, not only to the early days of the twentieth century, but also to the origins of the Universe. I hope that it is a journey that you enjoy.

2

The neutrino hypothesis

It is difficult to find a case where the word 'intuition' characterises a human achievement better than in the case of the neutrino invention by Pauli.[1]

Bruno Pontecorvo, 1980.

'Dear radioactive ladies and gentlemen . . .' Thus begins one of the most famous letters of modern physics. The writer was Wolfgang Pauli, a 30-year old professor of theoretical physics at the Federal Institute of Technology in Zürich; the date was 4 December, 1930; and one of the principal people to whom Pauli addressed his letter was Lise Meitner, a fellow Austrian.

Since 1907, Meitner had been working in Berlin on experimental studies of radioactive materials. In December 1930, she was among the experts on radioactivity who had gathered for a meeting in Tübingen – a meeting that Pauli declined to attend because, as his letter explains, 'a ball which takes place in Zürich the night of the sixth to seventh of December makes my presence here indispensable'.[2] However delightful this excuse, it was not the main message in the letter. Rather, Pauli had decided to share with his scientific colleagues 'a desperate way out' of certain paradoxes that had arisen in the nascent subject of nuclear physics.

The 'desperate way out' was the suggestion of a new subatomic particle – the particle that became known as the neutrino. At the time physicists knew of only three varieties of what we now call 'subatomic particles': the electron, the proton and the 'particle of light', the photon. In this context, Pauli's idea did indeed seem radical, for it would greatly modify this simple picture. But Pauli considered the problems in nuclear physics serious enough to warrant a desperate remedy, and so he turned to his 'radioactive' colleagues:

For the time being I dare not publish anything about this idea and address myself confidentially first to you, dear radioactive ones, with the question how it would be with the experimental proof of such a [particle].[3]

The crisis in nuclear physics that drove Pauli to such daring tempered by caution had come to a head in the closing years of the 1920s. We now know that there were several separate problems, and that the neutrino was to solve only one of these. But this particular problem had been a source of controversy for more than 20 years. It concerned the phenomenon of radioactivity, and one of the key figures involved in the controversy had been Lise Meitner.

The trouble with beta-rays

I am at present trying to write up the subject of beta rays for my new edition, and I find it the most difficult task in the book . . .[4]
Ernest Rutherford in a letter to Otto Hahn, 1911.

The French physicist, Henri Becquerel, discovered radioactivity in 1896. He found that photographic plates that he had left for some days in a drawer together with some uranium salts had become 'fogged', as if exposed to light. His conclusion, confirmed in subsequent studies, was that the uranium salts emitted some form of radiation, which Becquerel called 'les rayons uraniques' – uranium rays.

Many scientists quickly took up the opportunity of studying this new phenomenon, among them Ernest Rutherford, a 25-year old New Zealander who had won a scholarship to Cambridge University. In 1896, Rutherford was working in the Cavendish Laboratory under the director, Professor

These blurred images were formed on a photographic plate that Henri Becquerel had stored in a drawer under some uranium salts in February 1896. This was his first evidence for radio-activity. (AIP Niels Bohr Library, William G. Myers Collection.)

Ernest Rutherford came from New Zealand to England in 1895 to study at Cambridge University – he is pictured here (front right) with fellow research students at the Cavendish Laboratory in 1897. During his systematic investigations of 'uranium rays' in 1897–8, he discovered two components of differing penetration. He called those that were most easily absorbed 'alpha-rays', and those that penetrated less readily 'beta-rays'. (University of Cambridge, Cavendish Laboratory, Madingley Road, Cambridge.)

Joseph John ('J.J.') Thomson. Together, they were studying how X-rays ionise gases, in other words how the rays split the gas into positively and negatively charged components. Becquerel had shown that the uranium rays had a similar effect on travelling through air, so it was only natural that Rutherford should soon turn to study the new rays.

During 1897–8, Rutherford embarked on a systematic study of the absorption of the uranium rays, and in the course of this work he made an important discovery. The rays appeared to have two components. One component, which Rutherford called alpha-radiation, was easily absorbed by an aluminium foil no thicker than 0.002 cm. A second component was, however, some hundred times more penetrating. Rutherford called this beta-radiation. (A French physicist, Paul Villard, discovered a still more penetrating radiation in 1900; this third type of radiation naturally became known as gamma-radiation.)

Rutherford himself performed many of the crucial experiments that led to the identification of the alpha-rays, a crusade that took almost ten years. It was not until 1908 that with his colleague, Hans Geiger, he could write that 'we may conclude that *an alpha-particle is a helium atom*, or, to be more precise, *the alpha-particle, after it has lost its positive charge, is a helium atom*'[5] (their italics). But it was still not possible for them to take the next and final step of identifying alpha-particles as helium *nuclei*. Only late in 1910 did Rutherford make the crucial interpretation of results that Geiger and Ernest Marsden had obtained, which led to the concept of the atomic nucleus.

The identification of beta-radiation proceeded more quickly, with researchers in several laboratories making important contributions (although not Rutherford!). For example, in 1900 Pierre and Marie Curie – famous for purifying radium from pitchblende – measured the electric charge of the rays, which they found to be negative. Around the same time, Becquerel measured the ratio of electric charge to mass for the rays, and he found it to be similar to the value for cathode rays – in other words, electrons. Then in 1902, Walter Kaufmann at Göttingen reported on his decisive experiments that showed once and for all that beta-rays are streams of electrons.

At this time, there was no debate as to where in the atom the beta-ray electrons originated. Physicists had known since experiments by Kaufmann and by J.J. Thomson in 1897 that atoms contain electrons, and in 1902 Rutherford's model of the atom, with electrons orbiting a compact central nucleus, lay in the future. However, there was still the question of whether all the beta-rays emerging from a particular radioactive species have the same energy. This is certainly true for alpha-rays, so why not for beta-rays also? This was the question that Otto Hahn and Lise Meitner decided to investigate in 1907.

The 28-year old Hahn had recently joined the Chemistry Institute at the University of Berlin, after a short spell working with Rutherford, who was now professor at McGill University in Montreal. Meitner, only a few months older than Hahn, had studied physics in Vienna, where she had done some work on the absorption of alpha- and beta-rays. In 1907, she went to Berlin, where she soon encountered Hahn, and so began a partnership in research into radioactivity that was to last more than 30 years, until broken apart by the divisive politics of Germany in 1938. It was a distinguished partnership, which gave birth to several new radioactive materials and led ultimately to the discovery of nuclear fission – the phenomenon that underlies the workings of nuclear reactors and the original 'atomic' bomb.

At the time that Hahn and Meitner decided to work together, the director of the Chemistry Institute, Emil Fischer, did not allow women in the institute. He told Meitner that he had worried constantly about a Russian

Lise Meitner and Otto Hahn worked together for 30 years on research into radioactivity, until Meitner was driven from Germany by the Nazis. They are best known for their work leading to the discovery of nuclear fission, but some of their earliest research, beginning in 1907, was to investigate the energy spectrum of electrons emitted as beta-radiation from radioactive materials. (Ullstein Bilderdienst.)

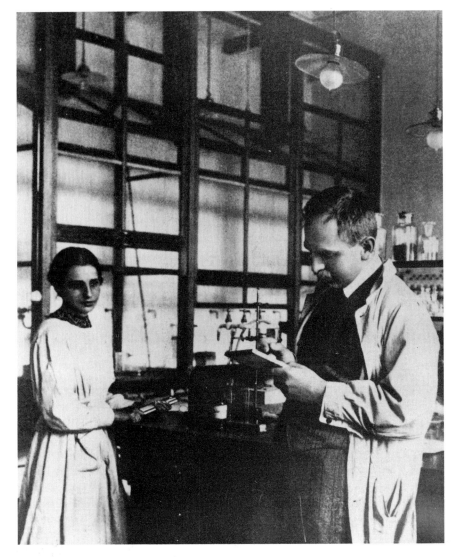

student 'lest her rather exotic hairstyle result in its catching fire on the Bunsen burner.'[6] So, in 1907, Hahn and Meitner started work in a converted carpenter's workshop in the institute, although two years later Meitner was allowed into the chemistry department.

They had decided to investigate the hypothesis that pure radioactive substances emit beta-rays with a single velocity, or equivalently, energy. At first, they believed that their findings supported this view, and in 1908 published results to this effect. But it turned out that they had been misled in their assumptions about the way in which electrons are absorbed in a material. The absorption they measured in successive thicknesses of aluminium did not, in fact, reflect a unique energy for the beta-rays. So,

Hahn and Meitner turned to a different technique, working with Otto von
Baeyer from the Physics Institute, a little way up the road.

Unlike the Chemistry Institute, the Physics Institute had magnets at its
disposal, and von Baeyer was able to use one of these to build a simple beta-
ray spectrometer. In this device, the beta-ray electrons, which are negatively
charged, travel on curved paths as they pass through a magnetic field. The
greater the velocity (that is, the energy) of the electrons, the less
the curvature. In this way, the magnetic field spreads the rays out in space
according to their velocity, rather as a prism spreads out a beam of light
according to its colour.

Together with von Baeyer, Hahn and Meitner used the spectrometer to
study the velocities of beta-rays from the purest, thinnest samples they could
make. The thinness was important to reduce effects that might be induced as
the beta-rays travelled through material before emerging into the magnetic
field. They prepared the samples at the Chemistry Institute and then took
them, as quickly as possible, to the Physics Institute. As Meitner later
recalled:

Hahn and I attempted to precipitate in as radioactively pure a condition as possible
the substances whose beta radiation we wished to investigate in the thinnest possible
layers on very short lengths of very thin wire . . . if our efforts were successful, we
raced out of the Chemistry Institute as if shot from a gun, up the road to the Physics
Institute a kilometre away, to examine the specimens in von Baeyer's very simple
beta spectrometer.[7]

The results of all this effort led to the conclusion, in 1911, that the beta-
rays from a pure substance have various energies, although the idea of a
single energy was not quite dead. Hahn and Meitner clung to the view that
the electrons were emitted with a single energy, but that the velocities
became modified by some 'secondary cause'. Indeed, by 1922, the picture
had become more complex, as several researchers began to find evidence for
many 'spectral lines' – in other words, clusters of beta-rays at various specific
energies.

Most of these studies relied on basically the same technique to study the
energy spectrum of beta-rays: a spectrometer to spread the rays out

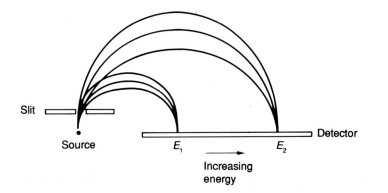

Slit

Source E_1 E_2 Detector

Increasing
energy

In a simple beta-ray spectrometer, a magnetic field bends the
paths of electrons from a
radioactive source as they travel
towards a detector. The electrons
follow different paths, depending
on their energy, the most energetic being bent the least by the
magnetic field. The result is that
electrons of different energies
arrive at different points on the
detector, which in the early work
of Meitner and Rutherford,
among others, was a photographic plate. (The magnetic field
in this sketch is perpendicular to
the page.)

according to energy, and a photographic plate to record the intensity of rays bent through different angles. But early in 1914, Rutherford received a letter referring to work with a slightly different technique:

I get photographs very quickly easily, but with the counter I can't even find the ghost of a line. There is probably some silly mistake somewhere.[8]

The letter was from James Chadwick, and the counter he referred to consisted of a metal plate and a sharply pointed needle.

Chadwick, who originated from Bollington in Cheshire, had studied physics at Manchester University, where Rutherford was professor from 1907 to 1919. Inspired by Rutherford, Chadwick stayed at Manchester after graduating in 1911, to continue research in physics. Then in 1913, he won a scholarship, which stipulated that he must move somewhere new to work. It was natural that Chadwick should choose to go to Berlin, to join Hans Geiger at the Physikalisch Technische Reichsanstalt. Geiger had previously worked several years with Rutherford at Manchester on studies of the nature of alpha-particles and on the famous alpha-particle scattering experiments that led to Rutherford's discovery of the atomic nucleus.

Once in Berlin, Chadwick attempted to use Geiger's 'point counter' to count the number of electrons contributing to each line in the beta-ray energy spectrum. In this counter, an electric field was established between a metal plate and a needle. When charged particles, such as beta-rays, passed through the counter they ionised the air, splitting it into positive and negative components. The effect was to discharge the electric field in the counter, the size of the discharge depending on the number of particles.

Chadwick used a narrow slit to select beta-rays emerging from the spectrometer within only a small range of the whole energy spectrum, and he measured the intensity of radiation passing through the slit with the point counter. Then, by varying the magnetic field in the spectrometer he could select different parts of the spectrum, that is, different energies. By April 1914, after a thorough study, he had convinced himself that the energy spectrum of beta-radiation, far from being a complex line-spectrum, is basically a smooth, continuous spectrum covering a broad range of energies. This is superimposed with only a few lines.

How could this be? Very competent scientists, from Hahn and Meitner to Rutherford himself, had claimed to observe many lines in the spectra of beta-radiation from various materials. Chadwick realised the explanation and published it along with his results. The photographic technique, which other researchers had used for detecting the beta-particles, was misleading: a small change in the number of particles could produce a large difference in intensity on the developed plate. And the final image depended on the development times. Thus Chadwick explained how at energies where the intensity was only slightly greater due to an additional line, the final photograph yielded dark lines against an apparently white background.

Thus ended Part I of the saga of the beta-ray spectrum. Part II was to begin only after the First World War, in which many scientists became

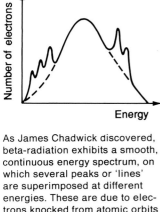

As James Chadwick discovered, beta-radiation exhibits a smooth, continuous energy spectrum, on which several peaks or 'lines' are superimposed at different energies. These are due to electrons knocked from atomic orbits either by electrons in the beta-radiation, or by gamma-rays emitted by the radioactive nucleus.

involved in different work. Meitner became an X-ray nurse with the Austrian army; Hahn, a reservist, joined an infantry regiment in the Germany army, and later became involved in the chemical warfare project; and Chadwick, who had been immersed in his studies of beta-rays in Berlin when war broke out, was interned at Ruhleben, a camp based in the stables of a racecourse near Spandau.

The dénouement

> . . . the long controversy about the origin of the continuous spectrum of beta-rays
> appears to be settled.[9]
>
> *Charles Ellis and William Wooster, 1927.*

A scientific society formed in the internment camp at Ruhleben, and here Chadwick met Charles Drummond Ellis – a cadet from the Royal Military Academy in Woolwich, who had gone for a holiday in Germany in the summer of 1914 and was still there when war broke out. The experience diverted Ellis from a future career in the army to one in physics, and after the war he soon enrolled at Trinity College, Cambridge. He was to play a major role in the conclusion of the beta-ray controversy in the 1920s.

When the physicists returned to their normal research in the wake of the war, the study of beta-rays had become part of a new field of discovery – nuclear physics. Rutherford had established the existence of a small but massive positive nucleus at the heart of the atom in 1911. And in 1913, the Danish theorist Niels Bohr had published the first quantum theory of the atom, based in part on Rutherford's discovery.

According to Bohr's theory, electrons in an atom can take up only certain specific energies. That is, the energy of the electron is *quantised*, each allowed energy being associated with an allowed orbit around the central nucleus. Bohr soon realised that the beta-ray electrons could not possibly originate from the electrons orbiting the nucleus: the energies of the beta-rays are simply too high. The most obvious alternative was that they must originate in the nucleus, implying that the nucleus must contain electrons.

During the next few years, Rutherford continued his studies of the nucleus and found that it must at least contain positively-charged particles with a charge equal in size but opposite in sign to that of an electron. These particles, which Rutherford called *protons*, occur singly as the nuclei of hydrogen atoms, but are clustered into larger groups in the nuclei of all other elements.

The nucleus could not be made from protons alone, however. In any atom, the total positive charge of the nucleus has to balance the total negative charge of the electrons, making the atom neutral overall. But a nucleus that consists simply of the same number of protons as there are electrons turns out to be too light, by more or less a factor of two. This observation led to the conclusion that the nucleus must contain protons *and* electrons – the negative charge of the nuclear electrons counteracting the positive charge of the additional protons required to make up the weight of the nucleus. After all, in beta-radiation electrons emerge from the nucleus, so what nuclear model could be more natural?

The electron-proton model of the nucleus was firmly entrenched by the early 1920s. In his popular science book *The ABC of atoms*, published in 1923, philosopher Bertrand Russell states: 'Everything known about nuclei is consistent with the hypothesis that they are composed of hydrogen nuclei [that is, protons] and electrons.'[10] It was against this background that the problem of the beta-spectrum resurfaced.

Ellis, inspired by Rutherford who was now director of the Cavendish Laboratory at Cambridge, had begun a series of important experiments. He was investigating the production of electrons from materials bombarded by gamma-rays. (Gamma-rays are the high-energy photons that constitute a third type of emission from radioactive materials.) In the course of his work, Ellis came to an important conclusion: that some of the lines in the beta-spectrum are due to electrons knocked from the inner atomic orbits by gamma-rays from the nucleus *of the same atom*. This is the process now known as *internal conversion* and it goes part way to explaining the existence of the lines superimposed on the continuous spectrum.

Ellis published these results in 1921. Meitner, who was now working at the Kaiser Wilhelm Institute for Chemistry in Dahlem, Berlin, fired the next salvo the following year. She put forward a complex explanation for beta-ray energies, still based on the assumption that all the electrons emerged from the nucleus with the same energy. Moreover, she ignored entirely the continuous component of the spectrum, which Chadwick had discovered. This was apparently because she thought that Chadwick's technique was not capable of resolving individual lines. Ellis immediately set to work, repeating Meitner's measurements, on which she had based her hypothesis, and found discrepancies. His conclusion: there was no evidence for Meitner's theory.

The controversy rumbled along, and in 1925 Ellis and William Wooster started an experiment that they hoped would settle the matter. Their aim was to measure the total energy associated with a single beta-decay. If the beta-electrons started out with a unique energy, which became distributed among secondary particles as Meitner proposed, the total energy of all the products would always add up to the initial amount. It would also presumably correspond to the maximum energy of the observed spectrum. If, on the other hand, the beta-electrons emerged with a whole range of energies, then after numerous decays, Ellis and Wooster would measure only the average value.

Their idea was to build a suitable calorimeter: a container that would absorb all the energy due to beta-decays, and which would show a rise in temperature as a result. They used a tube of lead, 13 mm long and 3.5 mm diameter, with a narrow central hole, little more than 1 mm in diameter. Their radioactive source consisted of radium-E (now known as bismuth-

Several key players in studying the spectrum of beta-radiation were among the participants at the Seventh Solvay Conference in Brussels in October 1933. Rutherford had discovered beta-radiation; Meitner had observed lines in the spectrum; Chadwick, on the other hand, found that it appeared smoothly varying. Charles Ellis eventually proved that the spectrum *is* continuous, but it required the intuition of Wolfgang Pauli eventually to explain the reason why. (Institut International de Physique Solvay, courtesy AIP Niels Bohr Library.)

210), deposited electrolytically in a thin coating on a platinum wire, which they placed in a brass tube that would fit just inside the lead tube. The brass absorbed alpha-rays from polonium, the daughter product of the radium-E.

The experiment was tricky; the temperature rises involved were only a thousandth of a degree centigrade or so. But the results were conclusive. Ellis and Wooster measured an energy per decay of 0.35 MeV (million electronvolts). This was close to the average value for the beta-spectrum from radium-E, and well below its maximum value of around 1 MeV.

(The electronvolt (eV) is the basic unit used to express energies in atomic and nuclear physics. One electronvolt is the energy an electron gains when it passes across the terminals of a 1-volt battery. On a macroscopic scale, relevant to everyday life, this is an unimaginably tiny amount of energy, but at the atomic level it is a natural unit. However, the higher energies associated with nuclear processes are more easily expressed in the larger units of millions of electronvolts, or MeV.)

Meitner was astounded. After she heard the news, she set about repeating the experiment, together with Wilhelm Orthmann. In December 1929, two years after Ellis and Wooster had published their results, Meitner and Orthmann had confirmed them. The energy spectrum of beta-radiation really was continuous.

The energy crisis

> Lately I have been thinking a good deal of the possible limitation of the conservation theorems . . . and . . . if in the reversal of beta-ray transformations we might find the mysterious source of energy . . . of stars.[11]
> *Niels Bohr in a letter to Ralph Fowler, 1929.*

It would be wrong indeed to create the impression that Meitner, Hahn and the others were being simply stubborn in clinging to the notion of unique energies for the emission of beta-rays. On the contrary, they held to an idea that is fundamental to many physicists: the concept that nature is simple, and should not require a great variety of theories in order to be understood. They were also guided in a way that often arises in scientific research – by analogy.

Research had shown that alpha-rays are usually emitted all with the same energy, or if not, at least in separate groups, each with a well-defined energy. Moreover, Bohr's quantum theory of the atom had shown how atomic phenomena could be understood only if you abandoned the idea that nature should behave continuously. The success of Bohr's theory lay in taking the counter-intuitive step of assuming that processes within the atom occur in jumps, each process or 'jump' being associated with a specific change in energy. It was therefore surprising to encounter a subatomic process – the emission of beta-rays – that appeared to have a continuous nature.

There was, however, a more serious problem with the continuous energy spectrum for beta-rays. It seemed a violate a well-respected principle of physics – the conservation of energy. This 'law', which is based on a wealth of experimental evidence, asserts that the total energy of a system must remain unchanged, unless some outside agency acts upon it.

What, in the late 1920s, did this law imply for the emission of beta-rays? The arguments went along the following lines. In beta-decay, an atomic nucleus emits an electron. In so doing, the nucleus changes – it loses one unit of negative electric charge, and becomes the nucleus of a different element. It also loses energy. The total energy before the decay is the energy equivalent to the mass of the original nucleus. (Einstein showed the equivalence of mass and energy in his special theory of relativity in 1905). But the new nucleus produced by the decay is always lighter than the original nucleus – and by a greater amount than can be simply accounted for by the mass-energy of the emitted electron. The reasonable assumption is that the missing mass-energy has been converted into kinetic energy, the energy of motion of the electron.

The amount of energy released should be the same in all decays of a particular type of nucleus, as the nucleus will always create the same decay product. This implies that the electron must always emerge with the same energy – contrary entirely to the observations of Chadwick, Ellis and Wooster!

This, then, was the problem with beta-decay, and one person who took it to heart was Niels Bohr. His proposed solution, which he had probably begun to consider late in 1928, was revolutionary: to abandon conservation of energy within the atomic nucleus.

On 8 May, 1930, Bohr gave the Faraday Lecture to Fellows of the Chemical Society at the Salter's Hall, in London, where he spelt out much of his thinking about the current problems with nuclear physics. He elaborated upon the theme in an extended script of the lecture, which was not published until 1932, when it appeared in the *Journal of the Chemical Society*. Here Bohr asserts:

At the present stage of atomic theory, however, we may say that we have no argument, either empirical or theoretical, for upholding the energy principle in the case of beta-ray disintegrations, and are even led to complications and difficulties in trying to do so.[12]

Nevertheless, he realises the seriousness of this proposition and adds:

Of course, a radical departure from this principle would imply strange consequences . . .[13]

Bohr was flying in the face of the accepted principles of physics indeed – not that he was a stranger to bold ideas, for his quantum theory of the atom had been based on dramatic new principles. Later in the Faraday lecture, he makes this very point:

Still, just as the account of those aspects of atomic constitution essential for the explanation of the ordinary physical and chemical properties of matter implies a renunciation of the classical idea of causality, the features of atomic stability, still deeper-lying . . . may force us to renounce the very idea of energy balance.[14]

It is only fair to point out that the energy problem of the continuous beta-ray spectrum was only one of the difficulties holding back progress in understanding the atomic nucleus. And it was only through a consideration of the whole messy scenario that Bohr was led to his radical proposals.

The very business of having electrons in the nucleus, as physicists then believed was the case, led immediately to a paradox. As Bohr explained, again in his Faraday Lecture:

... the present formulation of quantum mechanics ... is quite unable to explain why four protons and two electrons hold together to form a stable nucleus [helium]. Evidently we are here entirely beyond the scope of any formalism based on the assumption of point electrons, as ... the size of the helium nucleus, as deduced from the scattering of alpha-rays in helium, is of the same order of magnitude as the classical electron diameter.[15]

In other words, it was not possible within the bounds of quantum theory to fit electrons into a nucleus of the size that experiments indicated.

A third problem with the nucleus concerned the property of *spin*. This is an intrinsic angular momentum that both the electron and the proton possess. Two young Dutchmen, George Uhlenbeck and Samuel Goudsmit, discovered the spin of the electron in 1925. Bohr's theory of the atom had quantised the angular momentum of electrons in atoms – the momentum could have only specific values, which were associated with integer *quantum numbers*: 0, 1, 2, 3. . . . Goudsmit and Uhlenbeck found that they could explain certain details in atomic spectra if they introduced an additional quantum number for the electron, with a value of $\frac{1}{2}$. This implied an additional angular momentum, which the two researchers likened to spin, the two allowed values of the associated quantum number corresponding to clockwise $(+\frac{1}{2})$ and anticlockwise $(-\frac{1}{2})$ rotation. Two years later, David Dennision, at the University of Michigan, proposed that the proton is endowed with the same spin as the electron, that is, spin $\frac{1}{2}$.

The problem in the proton–electron model of the nucleus came with adding the spins of electrons and protons together within a nucleus. In particular, in 1929 a young Italian, Franco Rasetti, working at the California Institute of Technology in Pasadena, found that the nitrogen nucleus behaves as if it has a total spin of 1. But according to the proton–electron model, this nucleus should contain 14 protons and 7 electrons – in other words, a total of 21 subatomic particles, each with a spin of $\frac{1}{2}$. These spins could not possibly combine together to give a whole number. It was as if the electron somehow lost some of its usual characteristics, once inside the nucleus – a feature that Bohr referred to as the "remarkable 'passivity' of the intra-nuclear electrons".[16]

A desperate remedy

> In June 1931, on the occasion of a conference in Pasadena, I proposed the following interpretation: the conservation laws remain valid, the expulsion of beta particles being accompanied by a very penetrating radiation of neutral particles, which has not been observed so far.[17]
> *Wolfgang Pauli, Seventh Solvay Conference, 1933.*

On 1 July, 1929, Bohr wrote to his good friend and colleague at the Physics Institute in Zürich, Wolfgang Pauli. He enclosed two 'notes' concerning some problems he had been considering. One was a 'little piece about the beta-ray spectra',[18] which Bohr was unsure about publishing. He asked for Pauli's opinion, 'no matter how severe . . .'.

Pauli replied on 17 July with characteristic forthrightness:

Niels Bohr (right), discussed abandoning conservation of energy at the quantum level to explain the continuous spectrum of beta-radiation, but Pauli (left) argued that such a drastic move could not possibly be correct. (The Niels Bohr Archive.)

I must say that [the note about the beta-rays] gave me very *little* satisfaction . . . In Zürich Miss Meitner gave us a beautiful lecture . . . and she *almost* convinced me that the continuous beta-ray spectrum *cannot* be explained by secondary processes (gamma-ray emission etc.). So we really *don't* know what is the matter here. You don't know either . . . *In any case let this note rest for a good long time and let the stars shine in peace!*[19]

In this letter, Pauli expressed his distaste for Bohr's 'solution' to the beta-spectrum problem, but gave no alternative explanation. However, the seeds were sown and by December Pauli had put together his own scenario for beta-decay – although he was not keen to publish his ideas. Instead, he aired them in the famous letter to Meitner and other physicists assembled in Tübingen – the first written description of the particle now known as the neutrino.

Pauli referred to his new particles as 'neutrons' – *neutronen* in German. In the letter he says:

. . . there could exist in the nucleus electrically neutral particles, which I shall call neutrons, which have spin ½ and satisfy the exclusion principle and which are further distinct from light-quanta in that they do not move with light velocity. The mass of the neutrons should be of the same order of magnitude as the electron mass and in any case not larger than 0.01 times the proton mass. – The continuous beta-spectrum would then become understandable from the assumption that in beta-decay a neutron is emitted along with the electron, in such a way that the sum of the energies of the neutron and the electron is constant.[20]

The letter indicates how the introduction of Pauli's *neutronen* affects the distribution of energy in beta-decay. In this scenario, the energy released in the decay must be shared between *three* particles – the electron, the recoiling nucleus, and the 'neutron'. This can happen in many different ways, provided that the momentum of the electron and 'neutron' together balances that of the nucleus – in other words, as Pauli said, 'the sum of the energies of the neutron and the electron is constant'. At one extreme, the electron can take all the kinetic energy and the 'neutron' none – this corresponds to the upper limit of the measured beta-ray energy spectrum. At the other extreme, the electron is left with no kinetic energy, while the 'neutron' escapes with it all.

Pauli regarded his remedy as a 'desperate way out' yet he was drawn to it because it seemed to solve not only the problem of the continuous beta-spectrum, but also the problem concerning the spin of the nucleus. Pauli was not rejecting the idea that nuclei contain electrons; instead, he was suggesting that they also contain his *neutronen*, which like electrons and protons would have spin ½. According to this picture, nitrogen, for example, would contain 14 protons, 7 electrons and 7 *neutronen* – 28 particles each with spin ½ – and these could combine to give a total spin of 1, as observed. (In this case, 26 of the spins would cancel, while two would add together.)

To be able to inhabit the nucleus and yet make little contribution to the mass, Pauli's *neutronen* had to be similar in mass to electrons. But was it possible that such particles could exist without ever having been detected? Pauli postulated that they must be at least as penetrating as gamma-rays, the most penetrating component of radioactivity. And that raised the intriguing question of how to prove the existence of particles that had evidently so far escaped detection.

Pauli remained cautious. He aired his idea publicly for the first time on 16 July, 1931, at a meeting in Pasadena organised by the American Physical Society and the American Association for the Advancement of Science. He did not feel sufficiently confident to have his lecture printed – although a report did appear in the *New York Times*. Pauli's *neutronen* remained a possibility rather than a solution. It was to require the discovery of a different particle, plus the meshing together of several fundamental new ideas, to make the neutrino such a good idea that it had to be right. But all this was to happen during the next two years.

Rebuilding the nucleus

> It is, of course, possible that the neutron may be an elementary particle. This view has little to recommend it at present . . .[21]
>
> *James Chadwick, 1932.*

The picture of the nucleus began to change early in 1932, when James Chadwick at the Cavendish Laboratory in Cambridge demonstrated the existence of a new neutral particle. But it was not Pauli's hypothesised particle, for it was much too heavy, with a mass more or less equal to that of a proton.

In 1920, Rutherford had proposed that a neutral component of the nucleus should exist, and Chadwick had in the intervening years made several unsuccessful searches for such an object. His final, successful assault began after reading of results from experiments by Irène and Frédérick Joliot-Curie in Paris, published early in 1932. Walther Bothe and Herbert Becker in Berlin, and H.C.Webster in Cambridge, had previously observed a penetrating radiation produced when beryllium is bombarded by alpha-particles from polonium. The discovery the Joliot-Curies had made was that this radiation could knock protons from hydrogen with great effect, imparting high energies to the protons.

Soon after reading of these results, Chadwick went on to show that the new radiation could also knock atoms from other materials, such as carbon and nitrogen, again with relatively high energies. However, Chadwick argued that the radiation could not possibly transfer such a large amount of energy if it consisted of high-energy gamma-rays, as the Joliot-Curies had surmised. Instead, as he wrote in a paper published on 27 February in the journal *Nature*, 'The difficulties disappear . . . if . . . the radiation consists of particles of mass 1 and charge 0, or neutrons.'[22] By mass 1, Chadwick meant a mass equal to that of the proton; these particles were clearly not Pauli's *neutronen*.

Like Pauli, Chadwick was cautious in springing a new particle upon the

James Chadwick, whose discovery of the neutron in 1932 set the stage for the gradual acceptance of Pauli's new particle, the neutrino. Two decades earlier he had found that the energy spectrum of beta-rays is continuous. (Godfrey Argent.)

world – he titled the paper 'Possible existence of a neutron'. In support of his argument, he raised the spectre of nonconservation of energy, by pointing out that the radiation could be gamma-rays only 'if the conservation of energy and momentum be relinquished at some point.'[23]

What bearing did all this have on Pauli's hypothesis? Gradually, during 1932, physicists began to rethink their model of the nucleus – in came the neutron (that is, Chadwick's particle, the particle we still call the neutron, today), and out, slowly, went the electron. And it began to become clear that more than one solution might be necessary to resolve all the difficulties with the nucleus.

In August, in a paper presented to the French Academy of Sciences, Dmitrij Iwanenko from the Physico-Technical Institute in Leningrad took the step that Chadwick had seemed reluctant to make, and stated:

We do not consider the neutron as consisting of an electron and a proton but as an *elementary particle* . . . [and] are obliged to treat neutrons as possessing spin $\frac{1}{2}$[24].

Iwanenko removed the electron completely from the nucleus of nitrogen, by replacing seven of the protons in the old electron–proton model with neutrons of spin $\frac{1}{2}$. In the new model, nitrogen consisted of seven protons and seven neutrons – giving an even total number of particles, and a nucleus that could have spin 1. Thus the neutron alone solved the problem of nuclear spin – and set Werner Heisenberg on the road towards an understanding of the forces at work in the nucleus.

Heisenberg, a German, was a close friend and colleague of Bohr and Pauli. In 1925, when he was only 24, he had been instrumental in laying down the foundations of quantum mechanics – the mathematical framework that underpins Bohr's quantum theory of the atom. Seven years later, in 1932, he turned to applying quantum mechanics to the atomic nucleus.

In dealing with the nucleus, Heisenberg settled for a compromise. His nucleus was built from protons and neutrons, but the neutrons were themselves composite particles, each consisting of a proton and an electron. Heisenberg was mistaken in this last respect, but it did allow him to describe an 'exchange interaction' that could bind the nucleus together. In thinking of the force between a proton and a neutron, he imagined two protons held together by the exchange of an electron – in other words, the electron would reside first with one proton, thereby making it a neutron, and then with the other. The situation would be analogous, he argued, to the hydrogen molecular ion, H_2^+, in which two hydrogen nuclei (each a simple proton) share a single electron.

Heisenberg's picture may have been flawed – we know today that neutrons are not tightly bound couples of protons and electrons – but some aspects proved to be of profound importance. His concept of a nuclear exchange interaction underlies modern theories of the nuclear force. And the basic idea that the proton and neutron are somehow slightly different aspects of the same nuclear particle became a vital part of our understanding of subatomic particles.

Left, Werner Heisenberg, together with Niels Bohr. In the 1920s, Heisenberg developed Bohr's theory of the quantum nature of the atom into the mathematically based theory of quantum mechanics. He turned later, around 1932, to consider the nature of the forces that hold the nucleus together. (AIP Niels Bohr Library, Weisskopf Collection. Photo by P. Ehrenfest Jr.)

A view of creation

We then had a complete theory of radiation. The whole of the Einstein theory [of radiation] then followed from quantum mechanics.[25]

Paul Dirac, 1980.

Heisenberg was not being unreasonable in thinking of neutrons as proton–electron conglomerates – far from it. The phenomenon of beta-decay made it difficult to dispel electrons entirely from the nucleus. How else could electrons appear in beta-decay if they did not already exist in the nucleus?

Dmitrij Iwanenko believed he knew how. In the paper presented in August 1932, in which he rid the nucleus of electrons, Iwanenko commented that 'the expulsion of a beta electron [is] like the birth of a new particle . . .'[26]. Today, when the creation of particles at accelerators is commonplace, this does not sound very profound, but in 1932 it reflected some of the latest ideas in quantum theory. After all, how can a particle be created?

Fortunately, theoretical physicists in the 1920s had already been able to consider a particle that is readily and abundantly created – the entity we call the photon. The concept of a 'particle' or 'quantum' of light had been evolving gradually since 1905, when Albert Einstein introduced the idea of light as packets of energy, or light quanta. Over the ensuing years, physicists began to realise that these quanta also possess momentum; and the ballistic, particle nature of light was ultimately demonstrated in experiments in the early 1920s by Arthur Compton at the University of Chicago. A little later, in 1926, Gilbert Lewis, a distinguished chemist at the University of California, Berkeley, decided that these objects needed a name, and writing in the journal *Nature* in 1926 he proposed the name *photon*.

Photons are created whenever an object emits electromagnetic radiation, whether it be radiating heat (invisible infrared photons) and light (visible photons) or the more energetic X-rays and gamma-rays. Conversely, photons disappear when matter absorbs radiation. The theory that describes these processes is called quantum electrodynamics, and it began life in the mid-1920s with the work of one man in particular, Paul (P.A.M.) Dirac.

Dirac was a brilliant mathematician who became interested in quantum theory in 1924, while a postgraduate student at Cambridge University. Almost immediately, his work began to attract the attention of the main proponents of quantum theory, and in 1926 he went to work at Niels Bohr's Institute in Copenhagen. It was there that he wrote the paper that marks the beginning of quantum electrodynamics.

Dirac's major advance in this paper was to apply quantum mechanics to the interaction between an atom and electromagnetic radiation, and to come to grips with the problem of how an atom emits or absorbs radiation. One important step that he took was to follow a proposal by Werner Heisenberg and two colleagues, Max Born and Pascual Jordan. In quantum theory, any oscillating system has a host of possible energies, each associated with a quantum state. Born, Heisenberg and Jordan had suggested that each of these states could be interpreted in terms of a specific number of 'quanta', or lumps of energy – the number in question being the relevant quantum number. This is a process that is known, rather confusingly, as 'second quantisation', although nothing is quantised twice.

With electromagnetic radiation, Dirac was dealing with an oscillating system, so he decided to apply 'second quantisation'. He interpreted the many quantum states associated with the electromagnetic radiation field in terms of the related numbers of quanta – that is, photons. So a change from one state to another would involve a change in the number of photons.

As Dirac later recalled:

One thus gets a complete harmony of the wave and corpuscular theories of light. One can treat light as composed of electromagnetic waves, each wave to be treated like an oscillator; alternatively one can treat light as composed of photons . . . each photon state corresponding to one of the oscillators of the electromagnetic field.[27]

In Dirac's theory, the interaction between an atom and electromagnetic radiation became the interaction between the atom and a multitude of photons. And a change in the energy of the atom could be seen as the emission (creation) or absorption (annihilation) of a photon. This was the

Paul (P.A.M.) Dirac – his theory for the absorption and emission of photons was to provide Fermi with a blueprint for describing the appearance of neutrinos in beta-decay. (AIP Niels Bohr Library.)

beginning of quantum field theory – a general approach that is now applied not only to the electromagnetic field but also to other fundamental fields that govern the behaviour of matter. It is also the last new concept we need to have encountered if we are to appreciate the theory that endowed the neutrino with respectability.

Enter the neutrino

> We have all of course also been very interested in Fermi's new paper . . . although I must confess that I don't yet feel fully convinced of the physical existence of the neutrino.[28]
>
> *Niels Bohr in a letter to Felix Bloch, 1934.*

In November 1926, as Dirac was working on his theory of creation, a distinguished committee at the University of Rome was deciding who should occupy the university's new chair of theoretical physics. One candidate in particular appeared to have 'exceptional qualities' – the 26-year old Enrico Fermi, professor at Florence – and the distinguished physicists apparently had little difficulty in making their choice, voting unanimously for Fermi to take the new chair.

So Fermi returned to Rome, the city of his birth, and set about collecting around him a small but brilliant group of young physicists. Twelve years later, Mussolini's policies compelled Fermi and his Jewish wife, Laura, to leave Rome for the US, where they settled and where Fermi performed the work for which he is perhaps best known – the production of the first man-made nuclear chain reaction.

Less well publicised, but arguably more profound, was a theory that Fermi devised while still at Rome, in 1933. This was a theory that truly consolidated the proton–neutron picture of the nucleus and which set the seal on the old problem of beta-decay; a theory that involved not only the proton and the electron, but also the neutron and Pauli's particle, the neutrino; a theory that involved not only quantum mechanics, but also Heisenberg's linking of the proton and neutron, and Dirac's view of creation.

Fermi's first main exposure to Pauli's hypothesised *neutronen* came at a nuclear physics conference in Rome in October 1931. Fermi evidently asked Pauli to speak at the meeting on the new theory, but Pauli was still cautious and talked only privately with Fermi. It was about this time that Fermi introduced the name 'neutrino' for 'a little neutral object'. The word soon became commonplace among the physicists at Rome and later spread to those elsewhere.

Then came the discovery of the neutron. In Rome, a brilliant but charismatic colleague of Fermi's, Ettore Majorana, had surmised immediately from the results of the Joliot-Curies that they had discovered the 'neutral proton', or neutron. He then set about developing a model of the nucleus containing protons and neutrons, with no electrons. But Majorana refused to publish his ideas or to give Fermi permission to promote them, and soon others, such as Dmitrij Iwanenko, produced similar theories. However, the seeds were sown in Fermi's mind.

They were to blossom in the following year, 1933, in the wake of the Seventh Solvay Conference in Brussels, held in October. The meeting was attended by most of the major figures in nuclear physics, including key players in the beta-decay saga – Bohr, Chadwick, Ellis, Meitner and Pauli – and the young gurus of quantum theory, Dirac and Heisenberg. Here were the people who provided the key elements for Fermi's theory of beta-decay; but it took Fermi's special genius to mould them together.

Iwanenko had seen the analogy between beta-decay and the radiation of photons in 1932, when he said that the electrons are *born* in the process, being otherwise 'really very analogous to absorbed photons' in that they do 'not possess any individuality'.[29] Francis Perrin, who was at the same conference, was on the same track. In December, in a paper presented to the French Academy of Sciences, he stated:

If the neutrino has zero intrinsic mass, one must also think that it does not preexist in atomic nuclei, and that it is created, like a photon is, at the time of emission.[30]

Enrici Fermi (left), Werner Heisenberg (centre) and Wolfgang Pauli at Lake Como in 1927. Six years later, Fermi wedded Heisenberg's ideas on nuclear interactions together with Pauli's hypothesised neutrino in a highly successful theory of beta-decay. (AIP Niels Bohr Library, Segrè Collection. Photo by F. D. Rasetti.)

Fermi had the same basic idea, but in his mind it developed into a neat and powerful theory. Like Perrin, he wrote a paper in December, which he sent to the British journal, *Nature*, but the editorial staff of *Nature* rejected the paper as being too remote from reality. Instead, a short note appeared in Italian in *La Ricerca Scientifica*, and in January 1934, he submitted a full account of the theory to the Italian journal *Nuovo Cimento* and to the German journal *Zeitschrift für Physik*: both journals published the paper. (This dual publication was in line with a strict policy that Fermi followed: the Fascist government required all scientific work to be published in Italian, but few scientists in other countries read Italian journals, so Fermi published important work in German as well as in Italian.)

This paper is one of the classics of physics. In it, Fermi does not shy away from the new ideas of nuclear physics, but grasps them and turns them into a powerful theoretical tool that reproduced the experimental data on beta-decay. Nearly 60 years later, it remains as good a tool as any for dealing with low-energy beta-decays.

Let me try to convey something of how Fermi achieved all this. In the opening paragraphs of the paper, he lays out clearly the problems encountered with the continuous beta-spectrum and with supposing that the nucleus contains electrons. Equally clearly, he states the solutions he will assume. He will follow Pauli's suggestion that a lightweight, neutral particle – the neutrino – is emitted along with the electron in beta-decay; and he will follow Heisenberg in assuming that the nucleus consists only of 'heavy' particles – protons and neutrons. He was clearly not afraid of new ideas.

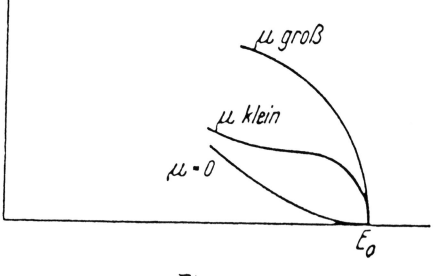

Figure 1 from Fermi's paper on his theory of beta-decay, 'Versuch einer Theorie der β-Strahlen. I', published in *Zeitschrift für Physik* **88**, 161, 1934. It shows how the calculated shape for the high-energy end of the electron spectrum varies with the hypothesised mass of the neutrino emitted at the same time as the electron. Measuring this shape forms the basis of many modern experiments to determine what mass the neutrino has, if any.

Fig. 1.

Then, Fermi describes some additional assumptions. First, he proposes that 'the total number of electrons and of neutrinos is not necessarily constant';[31] in other words, electrons and neutrinos can be created or can disappear, just like photons. Moreover, he states that '. . . to every transition from neutron to proton is correlated the creation of an electron and a neutrino . . . Note that by this the conservation of charge is assured.'[32]

There is one more important assumption. Fermi makes use of an 'internal coordinate' – a quantum number – for the heavy particles, which can take only one of two values: 1 for a neutron and − 1 for a proton. In this, he was following Heisenberg who had introduced the idea that protons and neutrons might be simply different quantum states of the same basic particle. (In modern theories, this quantum number has the value $\frac{1}{2}$ for protons and $-\frac{1}{2}$ for neutrons.)

Fermi's next step is to set up the equation describing the energy of the 'total system' – the heavy particles (the proton and the neutron), the light particles (the electrons and the neutrino) and the interaction between the heavy and light particles. The term for the interaction energy has to take account of the possible transformation of the neutron into a proton, accompanied by the creation of an electron and a neutrino; it has also to include the reverse process in which a proton turns into a neutron and an electron and neutrino are destroyed. Moreover, Fermi assumes that this interaction takes place at a single point in space – the position of the heavy particle.

Then comes the analogy with Dirac's radiation theory. In his theory, Dirac also had three terms in the energy equation – the energy of the atom, the energy of the radiation, and the interaction energy. He had treated the interaction as a perturbation on the other two components, for this allowed him to use approximations that made his calculations feasible.

Fermi follows Dirac as fully as he can. He takes the interaction as a perturbation on the energies of the heavy and light particles. Then, following standard methods, he arrives at an equation describing all possible transitions between the quantum states of the particles involved. This is the key to being predictive, and he is now in a position to calculate the probability for a transition to occur. What is more, he can reproduce the shape of the beta-spectrum.

The energy spectrum from beta-decays had dogged experimental physicists for more than 20 years, until it appeared definitely to be continuous. Now, two years after Pauli had hesitatingly introduced the neutrino into the world, Fermi could use his theory to tell something about the mass of the neutrino from the shape of the beta-spectrum near is high-energy limit. In his paper he shows three curves, for zero, small and large neutrino masses and comments:

The closest resemblance to the empirical curves is shown by the theoretical curve for [zero mass]. We come therefore to the conclusion that the rest mass of the neutrino is either zero or in any case very small with respect to the mass of the electron.[33]

The neutrino had arrived.

3

What is a neutrino?

Not everyone would be willing to say that he believes in the existence of the neutrino, but it is safe to say there is hardly one of us who is not served by the neutrino hypothesis as an aid in thinking about beta-decay.[1]

H. Richard Crane, 1948.

IF PAULI'S LETTER to the 'radioactive ladies and gentlemen' marked the conception of the neutrino, then Fermi's paper on beta-decay surely heralded its birth. Chadwick's discovery of the neutron had already removed some of the difficulties that Pauli's *neutronen* had originally been evoked to resolve. Now, in Fermi's theory, the role of a second neutral particle – the neutrino – became clear. Unlike the neutron, the neutrino is not something that resides in the nucleus, but is instead created at the instant of beta-decay. Other physicists began to consider the neutrino more seriously and very soon they were asking, how could you show the existence of neutrinos?

Subatomic particles are generally detected through their impact on their environment. Electrically charged particles – protons and electrons – directly ionise matter through which they pass; in other words, they knock electrons from atoms via electromagnetic interactions. In some detectors, the particles leave trails of ions and electrons behind them, and these can be made visible, like footprints in the snow. Other detectors register a burst of ionisation and thereby show that a particle has passed through. They can be used to count particles rather as a turnstile at a football stadium counts spectators.

Neutral particles – neutrons and photons – do not leave ionised trails in matter, but they can be detected indirectly. A photon can knock an electron from an atom, transmitting energy in the process so that the electron produces detectable ionised trails. Neutrons can knock protons out of nuclei, and the ionisation due to the charged protons then reveals the neutron's presence. Can neutrinos also be detected through some analogous effect?

Pauli, of course, had considered the question right from the start. He realised that if his *neutronen* were to exist then they must be more penetrating (in other words, less ionising) than gamma-rays, otherwise they would already have been detected in studies of beta-decay. But how much more penetrating?

Soon after the publication of Fermi's paper in 1934, James Chadwick at the Cavendish Laboratory in Cambridge, began investigating the neutrino's penetration, together with D.E. Lea. They positioned a solution containing radium-E about 6 cm from a detector and placed varying amount of lead, up to 5.8 cm thick, between the radium and the detector. They then measured the ionisation due to particles that penetrated through the shielding.

The radium would undergo beta-decay, emitting electrons, which would be absorbed in the lead. According to Pauli and Fermi, it should also emit neutrinos. The idea was that the neutrinos, slowed down by the lead, might now interact in some way to give rise, indirectly, to ionisation. But Chadwick and Lea found almost no ionisation beyond even the thickest lead shield, and they calculated that any neutral particles from beta-decay must on average be capable of travelling through at least 150 km of air before ionising a single atom. The following year, after performing a similar experiment 30 m below ground in Holborn underground station in London, M.E. Nahmias extended this distance to 31 000 km.

But could the neutrino interact at all with matter? In the wake of Fermi's paper, two German physicists, Hans Bethe and Rudolf Peierls, considered what seemed the only possibility – if a neutrino could be *created* in beta-decay, could one not be *absorbed* in the reverse process, in analogy to the creation and absorption of photons? Could a nucleus capture a neutrino to form a new nucleus and at the same time emit a negative or positive electron?

The positive electron, or *positron*, is another particle that had been discovered in 1932, following Chadwick's neutron by only a few months. It has the same mass as the electron, but its charge is positive – and its existence was explained immediately by theoretical work that Paul Dirac had published some five years prior to the discovery.

Dirac had brought together quantum theory and special relativity in his theory of the electron, developed at Cambridge during the winter of 1927–8. The theory led him to a famous equation, now known universally as the Dirac equation, which gives correctly the energy of an electron moving at 'relativistic' velocities, close to the speed of light. However, at first there appeared a drawback to the theory. It seemed to allow electrons to exist with negative values of energy, an apparent absurdity. But with the discovery of the positron it became clear that these positively-charged particles with real, positive energy could be associated with the negative-energy, negatively-charged particles that appeared in Dirac's theory. His theory of the electron was in fact a theory of electron and positron, or electron and 'antielectron'.

By early 1934, Irène and Frédérick Joliot-Curie, who had narrowly missed discovering the neutron, had found that positrons could be emitted in the radioactive decays of certain light nuclei, created artificially. This demonstrated that beta-decay could produce either negative or positive

In 1934, soon after Fermi published his theory of beta-decay, Hans Bethe and Rudolf Peierls calculated the extremely small probability for a low-energy neutrino to interact with matter. Here Bethe is seen in 1935 at Ann Arbor (left), while Peierls (right) is pictured on one side of Pauli with Dirac on the other. ((Bethe) AIP Niels Bohr Library, Goudsmit Collection; (Peierls) AIP Niels Bohr Library.)

electrons, depending on the nucleus involved. So, when Bethe and Peierls came to consider the absorption of a neutrino by a nucleus – which is often referred to as 'inverse' beta-decay – they assumed that either positrons or electrons could be emitted, again depending on the nature of the nucleus.

The two theorists found that the probability for neutrino absorption was far smaller than for any other process known at the time. In nuclear and particle physics, such probabilities are given in terms of areas or 'cross-sections' – the greater the probability that a reaction will occur, the larger the cross-section. Bethe and Peierls found 'what seemed . . . the fantastically small value of 10^{-44} cm^2,[2] and concluded that 'this meant that one obviously would never be able to see a neutrino'.

(This is an example of *exponential* notation, usually used to express very large or very small numbers. For example, 10^6 means 10 multiplied by itself 6 times, which is 1 000 000; and 10^{-6} means 1 divided by 1 000 000. So 10^{-44} means 1 divided by 10^{44}, or 1 divided by 1 followed by 44 noughts.)

Consider what the value of 10^{-44} cm^2 implies. A water molecule consists of one atom of oxygen and two atoms of hydrogen. Each hydrogen nucleus is

In the beta-decay of a neutron, the neutron emits an electron and an antineutrino as it turns into a proton. In inverse beta-decay, a proton absorbs an antineutrino and emits an antielectron, or positron, as it transmutes into a neutron.

Neutron beta-decay

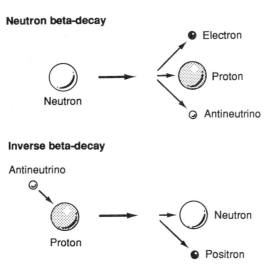

Inverse beta-decay

a single proton – a 'target' for neutrinos that we can imagine coming from a radioactive source. The probability that a neutrino will be absorbed by one of these target protons is given by the cross-section for the process multiplied by the number of targets in a unit of area one square centimetre:

Probability for reaction = cross-section × number of targets per unit area. In a cubic centimetre of water there are about 7×10^{22} 'free' protons – in other words, the nuclei of hydrogen, as opposed to the protons bound up with neutrons in the oxygen nuclei. According to the calculation of Bethe and Peierls, the probability for a neutrino to be absorbed by one of these protons is 10^{-44} multiplied by 7×10^{22}, which gives 7×10^{-22}. In other words, the chance that a neutrino is absorbed by free protons in a centimetre of water is about 7 in 10^{22}, or roughly 1 in 10^{21}. It would require 10^{21} cm of water to absorb a neutrino.

This is a length of astronomical proportions. It is about 1000 light years, or some 63 million times the distance from the Earth to the Sun. 'Obviously', as Bethe and Peierls concluded, no one would ever be able to detect a neutrino through inverse beta-decay. However, as Peierls commented almost 50 years later, he and Bethe did not allow for 'the existence of nuclear reactors producing neutrinos in vast quantities'[3] or for 'the ingenuity of experimentalists'. Nor did they realise that one day, physicists would have neutrinos with energies many thousands of times greater than those characteristic of nuclear reactions, and that such neutrinos would react much more readily with matter. But that story belongs to Chapters 4 and 5.

Neutrinos galore

I am not much impressed by the neutrino theory. In an ordinary way I might say that I do not believe in neutrinos . . . Dare I say that experimental physicists will not have sufficient ingenuity to make neutrinos?[4]

Arthur Eddington, 1939.

Even as these words of Arthur Eddington, a British astrophysicist of great renown, were being published in his book *The Philosophy of Physical Science*, the stage was being set for the production of enormous numbers of neutrinos. Late in 1938, Otto Hahn and Fritz Strassman in Berlin discovered that bombarding uranium with neutrons produced the much smaller nuclei of barium, which are about half the size of the uranium nucleus.

In the summer, following Hitler's invasion of Austria, Hahn's long-time colleague, Lise Meitner, had fled Germany. She was then nearly 60, but of Jewish descent, and her Austrian nationality no longer protected her from Nazi antisemitic laws. She found safety in Stockholm, where Hahn wrote to her with his amazing news, on 19 December. At Christmas, Meitner was joined by her nephew, Otto Frisch, also a physicist, and together they talked over the strange results. They decided that on absorbing neutrons, the uranium nuclei were becoming unstable, so that they split into two smaller, almost equal fragments. Meitner and Frisch named the process *fission*.

But the two lighter nuclei are not the only products of nuclear fission. As Meitner and Frisch realised, each fission would release a great deal of energy, on a scale millions of times greater than the energy liberated in the chemical reactions of normal combustion. Moreover, as Frédérick Joliot-Curie and his colleagues in Paris discovered four months later, several neutrons are also released by each fission. It was not long before physicists realised that these neutrons could themselves be made to initiate further fissions in a piece of uranium – in other words, a chain reaction could occur, allowing a truly enormous amount of energy to be produced from relatively small amounts of uranium.

The year was 1939. The Second World War broke out in September and the nuclear physicists soon found that their work was of interest to a far wider audience than usual. Three years later, Enrico Fermi and a team at the University of Chicago, backed by the US government, achieved the first man-made chain reaction on 2 December, 1942, in an 'atomic pile' built beneath the university's squash courts. This was the crucial step that set America on the road to building the world's first atomic bomb at a remote laboratory at Los Alamos in New Mexico. The bomb was exploded in the so-called 'Trinity test', on 16 July, 1945, in the New Mexican desert; within four weeks, the US had dropped atomic bombs on two Japanese cities, Hiroshima and Nagasaki.

In an atom bomb, the chain reaction proceeds rapidly, releasing the energy in a hugely destructive, explosive blast. By contrast, in nuclear reactors, the chain reaction is carefully controlled so that the energy is released steadily. But in both cases, another product of the uranium fissions escapes, barely noticed or even noticeable. For every fission, there emerge several neutrinos.

The neutrinos do not come directly from the fission reaction, but are instead produced by the fission fragments. These fragments are usually radioactive nuclei that transmute to more stable forms through a succession

A copious supply of neutrinos in the making – the world's first man-made nuclear reactor under construction at the University of Chicago in November 1942. (Argonne National Laboratory, courtesy AIP Niels Bohr Library.)

of beta-decays, each of which emits a neutrino. The result, from the sizable lumps of uranium in a bomb or a reactor, is a colossal number of neutrinos.

With the end of the war, the scientists who had worked on the American bomb project began to disperse and return to their pre-war studies. Some, however, stayed at Los Alamos, and in 1951, one of the physicists there found himself 'staring at a blank pad . . . searching for a meaningful question worthy of a life's work.'[5] Fred Reines, 33 years old, had been given leave to think of a suitable fundamental physics problem to study, aside from the deadly practicality of atomic bombs. Ideas were not too forthcoming and

Reines found that all that he 'could dredge up out of the subconscious was the possible utility of those bombs' neutrinos for direct detection.'[6]

This vague notion began to take on more substance after Reines found himself stranded at Kansas City airport with another physicist from Los Alamos, Clyde Cowan. In whiling away the time, the two fell to discussing the exciting questions of physics, and agreed that the detection of the neutrino represented a 'supreme challenge'. They decided to join forces to work out how to detect the neutrinos released during the testing of an atomic bomb.

Their basic objective was clear: 'to detect an interaction of the neutrino at a location remote from the point of origin'.[7] The 'obvious' interaction to choose was inverse beta-decay; more problematic was how to detect the interaction. In inverse beta-decay, a proton absorbs a neutrino and turns into a neutron, at the same time emitting a positron. The positron interacts almost immediately with a nearby electron, and the particle and antiparticle disappear into pure energy, in a process known as annihilation. The result is two gamma-rays, in other words, high-energy photons, which head off in opposite directions.

Cowan and Reines believed they knew how to detect the positrons via the gamma-rays and thought this might be sufficient; the neutrons produced would be more difficult to pin-point. (Strictly speaking, the general belief is that it is an *anti*neutrino that a proton absorbs in this reaction, as it is an *anti*electron that is produced. This is a topic we shall reconsider later in the chapter; for now I will continued to refer simply to 'neutrinos'.)

In 1950, several groups of researchers had discovered organic liquids that *scintillate*, or in other words produce small flashes of light almost immediately after a charged particle has passed through and ionised some of the atoms. This light can be converted to an electrical signal by photomultiplier tubes, devices that produce bursts of electrons when photons of light strike a special surface. As gamma-rays can induce ionisation indirectly, scintillations would mark the death of any positrons annihilating in scintillating material. The material could therefore act both as target, providing protons for the inverse beta-decay reactions, and as detector.

Reines and Cowan decided to use a tank of liquid scintillator viewed by several photomultiplier tubes, and to have it suspended below ground in a vertical shaft some 40 metres distant from the tower on which the bomb would be exploded. Air would be pumped from the shaft, so that once the bomb exploded the detector could be released to fall freely through a vacuum, unaffected by the shock-wave from the explosion. Once the shock-wave had passed, the detector would land on a bed of feathers and rubber foam. Now it could pick up the scintillations due to positrons emitted as neutrinos from the bomb's fission fragments flooded through the liquid. The surrounding earth would serve to absorb other kinds of particle fleeing the explosion.

It was an amazing proposal, but the two physicists were undaunted, as Reines recalls:

In their first scheme for detecting neutrinos, Fred Reines and Clyde Cowan planned to intercept some of the many billions of neutrinos emitted in the explosion of a nuclear bomb. (Smithsonian Institute, *Smithsonian Report for 1964*, p. 419.)

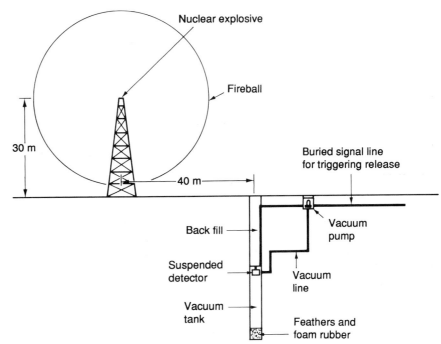

The idea that such a sensitive device could be operated in the close proximity (within a hundred metres) of the most violent explosion produced by man was somewhat bizarre, but we had worked with bombs and felt we could design an appropriate system.[8]

The confidence paid off – not because the experiment worked, however, as it was never put into practice. But in working through the proposal in more detail Cowan and Reines hit upon something that would allow them to make use of a less spectacular source of neutrinos – a nuclear reactor.

The neutron emitted along with the positron in inverse beta-decay, like any free neutron, is an elusive beast to capture. It can stagger through a material like a drunk in a crowd, colliding first with one nucleus and then with another, losing a little energy in each collision, until finally it is absorbed. Such an event is heralded when the nucleus that captured the neutron reorganises itself, emitting gamma-rays as it does so to rid itself of excess energy. Thus scintillations due to a second burst of gamma-rays could reveal the capture of the neutron, just as the first burst reveals the annihilation of the positron.

The crucial factor in using this technique in a viable neutrino detector turns out to be the neutron's drunken walk between is birth and its capture. This takes a characteristic average time, rather as there might be a characteristic length of time between a drunkard leaving a saloon bar and falling over in the street! So the two bursts of light due to inverse beta-decay in the scintillator are closely correlated in time: the flash due to the neutron must follow the one due to the positron by a specific time delay – a delay that can be calculated from a knowledge of the way neutrons interact with matter. For neutrons emitted in inverse beta-decay, travelling through liquid scintillator, this delay is about 5 microseconds.

Reines and Cowan realised that the delay between the positron and neutron flashes in the liquid scintillator provides a unique 'signature' for inverse beta-decay. By looking for pairs of flashes, separated by the right amount of time – about 5 microseconds – they could reject many of the unwanted flashes due to other particles reaching the detector, as these would occur at random intervals.

They were still working on their proposal to detect neutrinos from a bomb when they suddenly realised that they could do the experiment without a bomb – the characteristic delay would serve equally well to distinguish inverse beta-decays from unwanted reactions if their detector were located near a nuclear reactor.

Neutrinos seen

> Those days at Hanford were both stimulating and exhausting. For a few months we stacked and restacked several hundred tons of lead and boron-paraffin shielding. We worked around the clock as we struggled with dirty scintillator pipes, white reflecting paint that fell from the walls under the action of toluene based scintillator and cadmium proprionate neutron capturer . . .[9]
>
> *Fred Reines, 1982.*

Cowan and Reines set to work on their first neutrino detector at one of the reactors at the Hanford Engineering works in the state of Washington. Thus began 'Project Poltergeist', a name chosen to reflect the elusiveness of the neutrino. The detector itself consisted of a cylindrical tub holding about 300 litres of scintillator, viewed by 90 photomultiplier tubes, arranged around the walls of the tub. Lead and boron-paraffin surrounded the detector to shield it from the streams of neutrons and charged particles liberated in the fission reactions in the reactor's core. The cadmium propionate, on the other hand, was added to the liquid scintillator to improve the chance of capturing the neutrons released by inverse beta-decay in the scintillator.

The main difficulty was with 'background' – pairs of flashes with the correct time difference that occurred whether the reactor was on or off. The background seemed to be due mainly to cosmic-rays, including neutrons, which interacted during the 5-microsecond 'window'. (Cosmic-rays are penetrating subatomic particles that rain down through the Earth's atmosphere, and which arise when high-energy particles from outer space collide with atomic nuclei in the upper atmosphere.)

Calculations suggested that inverse beta-decay should lead to a pair of flashes with the appropriate 5-microsecond delay ('a delayed coincidence') about every 5–10 minutes, but the experiment detected about five coincidences a minute, irrespective of whether the reactor was on or off. However, there was a small, but discernible increase in the counting rate with the reactor on, which was of about the right amount.

(a) Clyde Cowan (far left) and Fred Reines (far right) with their team on 'Project Poltergeist', the prototype neutrino detector that demonstrated the potential of the technique they had chosen. (b) The detector itself – a 300-litre tank of liquid scintillator, surrounded by 90 phototubes. Before this, 20 litres of liquid had seemed a large volume! (Los Alamos National Laboratory.)

Reines and Cowan were encouraged. In a paper published late in 1953, they stated that: 'It appears probable that this aim [to detect the free neutrino] has been accomplished'.[10] But they recognised the need for more convincing evidence and set about designing a new detector which would be better at picking out the 'signal' of coincidences due to inverse beta-decay from the 'background' of random cosmic-ray reactions.

In the autumn of 1955, with the encouragement of John Wheeler, an eminent American theorist, Cowan and Reines, set up their improved detector, this time at a new, powerful reactor at Savannah River in South Carolina. Here they could shield their detector in an area 11 metres from the core of the reactor and 12 metres below ground – vital for reducing the troublesome background from cosmic-rays. Moreover, the more powerful reactor produced many more fissions each second than the device at Hanford, and so there were many more neutrinos arriving at the detector – more than 10 million million in every square centimetre each second.

In designing the new detector, Cowan and Reines took advantage of the fact that when a positron annihilates with an electron in matter, the two gamma-rays leave the 'scene of the crime' travelling in almost exactly opposite directions – a result of momentum conservation. And, when a nucleus captures a neutron, the gamma-rays emitted head away from the new nucleus. Detecting flashes from pairs of particles on opposite sides of the 'target' would therefore provide an additional means of discriminating against the unwanted, background reactions.

The final apparatus resembled a huge sandwich. It consisted of three tanks of liquid scintillator, interleaved by two 'target' tanks, filled with water. If a

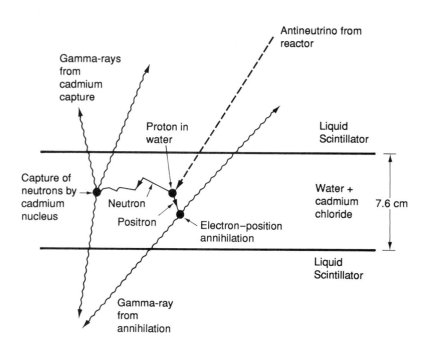

The method Reines and Cowan used finally to detect neutrinos was a three-step process: first a proton in the water would absorb a neutrino, and produce a neutron and a positron; then almost immediately the positron would annihilate with an electron, emitting two back-to-back gamma-rays which could be detected in the adjacent layers of scintillator; and lastly, some 5 microseconds or so after the annihilation gamma-rays had emerged, the neutron would be captured by a cadmium nucleus, which would then emit some more gamma-rays.

proton in the water should absorb a neutrino from the reactor, then the positron created would immediately give rise to two gamma-rays, through annihilation. The gamma-rays would speed off, back-to-back, into the two scintillator tanks, immediately above and below the target. In this way, the gamma-rays would produce a simultaneous flash in each tank, which would be picked up by photomultiplier tubes. Five microseconds later another coincident pair of flashes would appear in the same two scintillator tanks, this time from the gamma-rays radiated after the capture of the neutron. Cadmium chloride dissolved in the water would ensure the capture of the neutrons.

The apparatus weighed a total of about 10 tonnes, not including the shielding, and stood over 2 metres tall. Each scintillator tank contained 1400 litres of scintillator, which was viewed by 110 photomultiplier tubes. These tanks were steel, with thin lids and bases to allow low-energy gamma-rays to pass through. The target tanks were made from polyethylene and each contained 200 litres of water.

Cowan and Reines, aided by Francis Harrison, Herald Kruse and Austin McGuire, ran the detector for a total of 100 days, over about a year. During

The final assembly that Cowan and Reines used to detect neutrinos from the powerful nuclear reactor at Savannah River, South Carolina. The large tanks labelled I, II and III contained liquid scintillator, and were each viewed by 110 phototubes; the smaller ones sandwiched between, labelled A and B, contained water with cadmium chloride dissolved in it. (*Physical Review* **117**, 160, 1960.)

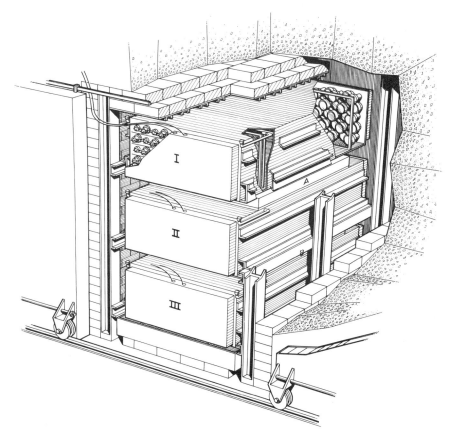

this time, when the reactor was on, they saw on average 3 'delayed coincidences' per hour – in other words, two pairs of flashes in two of the scintillator tanks, separated by the right amount of time. They performed many tests, checking for example that the first pair of pulses from the photomultiplier tubes looked like those from a positron source (copper-64) dissolved in the water during a test run; they doubled the amount of cadmium in the water to see if the time delay before the second pair of pulses decreased, as would be expected if the pulses were truly associated with neutron capture; they used water with no cadmium to show that they could no longer detect a signal when the reactor was on.

Eventually, after all the tests, they were convinced. As Reines recalls, 'It was a glorious feeling to have participated so intimately in learning a new thing'.[11] And on 14 June, 1956, he and Cowan sent a telegram to Pauli, to tell him of their success:

We are happy to inform you that we have definitely detected neutrinos from fission fragments by observing inverse beta-decay of protons. Observed cross-section agrees well with expected six times ten to minus forty-four square centimetres.[12]

The death of parity

> ... the symmetry between the left and the right, is as old as human civilization ... debated at length by philosophers of the past.[13]
>
> *Chen-Ning Yang, Nobel Lecture, 1957.*

Even as Cowan and Reines in Los Alamos were sending their telegram to Pauli, two young Chinese Americans in New York were in the throes of turning a cherished notion of physics on its head. Their ideas were to lead to the first significant alteration to Fermi's theory of beta-decay in more than 20 years. They dared to challenge the long-held belief that basic physical processes are symmetric between left and right.

On a 'human' scale, nature clearly distinguishes between right and left. Our hearts are generally displaced to the left side of our bodies, and many people find themselves able to manipulate one hand better than the other. But prior to 1957, most physicists had no reason to believe that, at a fundamental level, nature would distinguish between right and left.

Imagine then, their surprise when they discovered early in 1957 a phenomenon that pays no attention to this presumed spatial symmetry; an experiment that gives a different result when right and left are interchanged. More surprisingly still, the effect occurred in a well-known area of physics that had been studied for half a century – nuclear beta-decay.

In April 1956, Tsung-Dao ('T.D.') Lee, at New York's Columbia University, and Chen-Ning ('Frank') Yang from the Institute for Advanced Study in Princeton, New Jersey, had begun to look into some puzzles concerning the decays of so-called *strange* particles – short-lived particles discovered in the high-energy interactions of cosmic-rays. By early summer they had come to a remarkable conclusion. They realised that they could

Tsung-Dao Lee (left) and Chen-Ning Yang, who in 1956 proposed that interactions through the weak force are not left-right symmetric, as had long been assumed. A year later, they received the Nobel prize for their work. (AIP Niels Bohr Library, photo by Alan W. Richard.)

understand the puzzling effects *if* the particle decays did not respect left–right symmetry.

By the 1950s, research had shown that the decays of a number of subatomic particles proceed very much in the same manner as nuclear beta-decay. This led to the notion of a new force underlying the decays, rather as the electromagnetic force underlies the interactions of charged particles – a topic we shall return to in Chapter 4. The decays, and related reactions such as the absorption of neutrinos in inverse beta-decay, were very slow in comparison to electromagnetic interactions, so the new force became known as the 'weak force'. These developments provided Fermi's theory with a new role, as the first theory of the weak force, and it was in this context that Lee and Yang made their discovery.

They suspected that interactions via the weak force – exemplified in the strange-particle decays – did not respect left–right symmetry. But it was a big 'if'. Why should the weak force differ from other physical forces in this respect? And why had no one noticed this before? There was already a large body of data from experiments on beta-decay, the archetypal weak reaction, so Lee and Yang worked carefully through all the relevant results. They found no evidence for any 'violation' of spatial symmetry. Then, the reason hit them. None of the experiments had actually measured effects that depend on left–right symmetry in beta-decays!

A year later, when he accompanied Lee to Stockholm to receive the Nobel prize for their work, Yang described their feelings about their discovery:

The fact that [right–left symmetry] in the weak interactions was believed for so long without experimental evidence was very startling. But what was more startling was the prospect that a space time symmetry law which the physicists have learned so well may be violated. This prospect did not appeal to us. Rather we were, so to speak, driven to it through frustration with the various other efforts at understanding the [strange-particle] puzzle that had been made.[14]

On 22 June, 1956, Lee and Yang submitted their now famous paper on 'Question of Parity Conservation in Weak Interactions' to *The Physical Review*. 'Parity' is the term used to express mathematically the spatial symmetry properties of physical processes. It is said to be conserved in a process if the physical laws governing the process do not differentiate between left and right. In their paper, Lee and Yang wrote:

To decide unequivocally whether parity is conserved in weak interactions, one must perform an experiment to determine whether weak interactions differentiate the right from the left.[15]

Later, they suggested possible tests, beginning:

A relatively simple possibility is to measure the angular distribution of the electrons coming from beta-decays of oriented nuclei . . . an asymmetry of the distribution . . . constitutes an unequivocal proof that parity is not conserved in beta-decay.[16]

Yang and Lee were in close contact with a number of skilled experimenters. Yang often visited the Brookhaven National Laboratory on Long Island, while the department where Lee worked at Columbia University attracted many of the best physicists. A few floors above Lee's office was Chien-Shiung Wu, one of the acknowledged experts on beta-decay. She was fascinated by the theorists' proposals, and suggested an experiment using cobalt-60, a form of cobalt that undergoes beta-decays. Her idea was to polarise the nuclei in the cobalt – in other words, to align them in a magnetic field so that their axes of spin were all parallel. (Recall that many nuclei had been found to behave rather as if spinning as discussed briefly in Chapter 2.)

To polarise the cobalt-60 in this way, Wu needed to work at very low temperatures, only a fraction of a degree above the absolute zero of temperature. This would minimise the normal thermal jiggling of the nuclei, which would otherwise disturb the alignment. So Wu contacted Ernest Ambler and his team at the low-temperature laboratory of the Bureau of Standards in Washington. Together they began to devise an experiment to test left–right symmetry in beta-decays, an effort that was going to occupy them for much of the second half of 1956.

Chien-Shiung Wu, pictured in the laboratory at Columbia University in 1957, the year that began with the publication of her discovery that beta-decay is not left–right symmetric, and does indeed violate 'parity' or spatial symmetry. (AIP Niels Bohr Library.)

Then, on 29 December, Lee received a momentous telephone call from Wu:

She told me that her cobalt-60 result indeed showed parity nonconservation, but that further checks would still be necessary to ascertain the precise magnitude of the violation.[17]

Five days later, on 3 January, 1957, Wu appeared in Lee's office:

[She] said that she had checked all corrections and the violation effect was very large. I assured her that this was excellent . . . I then immediately called . . . Yang in Princeton.[18]

It had taken Wu, Ambler and their colleagues the best part of six months to set up the intricate experiment. But once it was working, they did not need many minutes to make their discovery. They found that when the spinning cobalt-60 nuclei were aligned, the beta-ray electrons no longer sprayed out equally in all directions. Instead, more of them emerged opposite to the direction of the magnetic field. When Ambler and Wu reversed the magnetic field, the direction of the electrons reversed also. This asymmetry was the 'unequivocal proof' that Lee and Yan had required. Parity conservation in beta-decays was dead, and the world of physics was astounded.

When he had first heard of the experiment being performed by Wu and Ambler, Wolfgang Pauli had written, in a letter to Victor Weisskopf, that he was willing to bet a large sum of money that the results would be symmetric. When news of the measured large asymmetry reached him, Pauli wrote again to Weisskopf:

Now the first shock is over and I begin to collect myself again . . . It is good that I did not make a bet. It would have resulted in a heavy loss of money (which I cannot afford); I did make a fool of myself, however (which I think I can afford to do) . . . the others now have the right to laugh at me.[19]

The vampire neutrino

On reflecting a neutrino in a mirror, one sees nothing.[20]
Abdus Salam, 1957.

In September 1956, three months before Wu and her colleagues observed parity violation for the first time, Yang gave a lecture at the Seattle International Conference on Theoretical Physics, entitled 'Present Knowledge About the New Particles'. Here he discussed the problem of the strange particles and raised the spectre of the nonconservation of parity. One member of the audience in particular found that the suggestion continued to haunt him as he flew back to London, courtesy of the American Air Force. Abdus Salam, from Pakistan, was working at the Cavendish Laboratory in England. He recalls:

. . . the plane was very uncomfortable, full of crying children of servicemen. I could not sleep. I kept reflecting on why nature should violate left–right symmetry in weak interactions.[21]

Salam believed that the key to all this – 'the hallmark of most weak interactions' – was the neutrino. He came back to a question he had been asked in his PhD examination, as to why the mass of the neutrino is apparently zero, and

during the comfortless night the answer came . . . Nature had the choice between a theory which is aesthetically satisfying but in which left–right symmetry is violated, with a neutrino which travels exactly with the velocity of light; and a theory where left–right symmetry is preserved, but the neutrino has a tiny mass . . .

It appeared at that time clear to me what choice nature must have made . . . I got off the plane the next morning, naturally very elated.[22]

At first Pauli, seen here with Wu in 1957, did not believe that parity could be violated by the weak force, but the results of Wu and her colleagues proved him wrong. (AIP Niels Bohr Library, Physics Today Collection.)

He had realised that when Dirac's equation for the electron is applied to *massless* particles, the theory violates left–right symmetry.

Salam rushed back to the Cavendish Laboratory in Cambridge and calculated various consequences before racing off again, this time to visit Rudolf Peierls at Birmingham, who had asked the question in the PhD exam. Peierls did not approve of the newly-discovered answer:

I do not believe left–right symmetry is violated in weak nuclear forces at all. I would not touch such ideas.[23]

Salam was later again discouraged, but with equal kindness, by Pauli:

Give my regards to my friend Salam and tell him to think of something better.[24]

Then came the stunning news from Wu and Ambler. Nature *did* seem to have chosen the option of parity-violating weak interactions.

Lee and Yang had also considered the implications of parity violation for the neutrino. On 10 January, less than two weeks after Wu's telephone call reporting the first indications of parity violation, they sent a paper on their new theory of the neutrino to *The Physical Review*. Like Salam, they realised that parity violation could arise from the Dirac equation for a neutrino of zero mass. They then simplified the way in which they represented the neutrino, using only two 'components' rather than the four needed for a massive particle, such as the electron. This immediately implied that the neutrino could exist in only two states – either as a neutrino spinning one way relative to its direction of motion, or as an *anti*neutrino spinning the other way. It was not possible that the neutrino (or antineutrino) could spin both ways.

This appeared to be the asymmetry at the heart of parity violation in weak interactions. Look at a neutrino in a mirror which interchanges left and right, and what should you see? A particle travelling in the same direction, with its spin reversed. But this cannot be a neutrino, for neutrinos can spin only one way relative to their motion. The neutrino is like the vampires of gothic fiction – it has no reflection!

The left-handed neutrino

A vague notion appeared to persist that a double definition could be used: the anti-neutrino, say, being defined either as the particle emitted together with a negative electron in beta-decay or as a left-handed screw. For a while it must have looked as if a definition had replaced a measurement![25]

Maurice Goldhaber, 1958.

A subatomic particle with spin can be assigned a 'handedness' – or *helicity* – rather as a corkscrew can. If it is spinning clockwise relative to its direction of motion, the particle is analogous to a right-hand screw thread. On the other hand, a particle spinning *anti*clockwise about its direction of motion is like a left-hand screw thread.

The 'two-component', massless neutrino theory suggested that neutrinos could be either left-handed or right-handed, but not both. But theory did

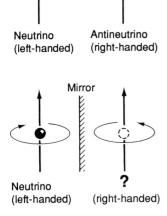

Neutrino
(left-handed)

Antineutrino
(right-handed)

Mirror

Neutrino
(left-handed)

?
(right-handed)

A neutrino spins one way about its direction of motion – anticlockwise, like a left-hand screw – while the antineutrino spins the other way. Reflect a neutrino in the parity 'mirror', however, and while the sense of spin reverses, the direction of motion does not. The reflected neutrino now has the wrong 'handedness', or helicity.

not say which way the neutrino emitted in beta-decay, for example, would spin. This was a matter for experiment to decide, and decide it did, at the Brookhaven National Laboratory.

Yang and Lee had spent the summer of 1956 at Brookhaven, and had talked frequently about their ideas on parity with experimental physicists there, in particular with Maurice Goldhaber, an experienced nuclear physicist. After the Nazis had come to power in 1933, he had interrupted his studies in Berlin to go to Cambridge. There he began research with James Chadwick at the Cavendish Laboratory, and showed for the first time that the neutron is heavier than the proton and must therefore be unstable. Goldhaber had then gone to the US in 1938, moving to the new Brookhaven National Laboratory on Long Island, New York, in 1950.

In the autumn of 1957, Goldhaber and his colleagues Lee Grodzins and Andrew Sunyar set up an experiment to measure the helicity, or handedness, of the neutrino. It was a *tour de force*, which even today draws admiration from physicists.

The idea was to study the process of electron capture, which is closely related to beta-decay. Certain unstable nuclei can capture a nearby electron from an atomic orbit. A neutrino then emerges, as a proton inside the nucleus changes into a neutron. In this way, the nucleus transmutes from one chemical element to another, with one proton less. The neutrino emerges spinning one way, while the nucleus recoils with back-spin. Conservation of both linear and angular momentum dictates that the nucleus and the neutrino not only head off in opposite directions but also spin in opposite directions. This, in fact, means that they have the same helicity. So, Goldhaber and his colleagues argued, *if* they could measure the spin and direction of the recoiling nucleus, they would know the helicity of the neutrino.

But how to measure the nucleus? The answer lay in choosing a nucleus that was spinning too much after capturing the electron, and which would rid itself of the excess spin by quickly emitting a gamma-ray. The spin of the escaping gamma-ray would reflect the unwanted back-spin of the nucleus. The task facing the experimenters would then be to select gamma-rays that emerged exactly in the direction of the nuclear recoil and to measure their spin, or in other words, their polarisation. These gamma-rays would reveal the helicity of the nucleus created by electron capture and hence the helicity of the emitted neutrino.

The first problem was to select a nucleus that would go through the appropriate sequence of events. The only suitable starting nucleus they found was a radioactive form of a rare metallic element called europium – europium-152*m* to be precise, where the *m* indicates a form of short-lived stability, or 'meta-stability', and 152 indicates the total number of protons (63) and neutrons (89) in the nucleus. When europium-152*m* captures an electron, it changes into samarium-152 (62 protons and 90 neutrons). The samarium nucleus formed in this way has too much spin, but it very swiftly emits a gamma-ray to become a stable form of samarium.

In their elegant experiment to discover the helicity of the neutrino, Goldhaber, Grodzins and Sunyar measured the helicity of gamma-rays emitted by a nucleus recoiling after neutrino emission. The nucleus transmutes by capturing an atomic electron and emitting a neutrino; the new nucleus and the neutrino move away back-to-back to conserve momentum.

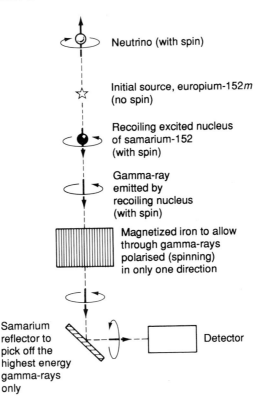

Neutrino (with spin)

Initial source, europium-152m (no spin)

Recoiling excited nucleus of samarium-152 (with spin)

Gamma-ray emitted by recoiling nucleus (with spin)

Magnetized iron to allow through gamma-rays polarised (spinning) in only one direction

Samarium reflector to pick off the highest energy gamma-rays only

Detector

Goldhaber, Grodzins and Sunyar could make the radioactive europium at the nuclear reactor at Brookhaven, by bombarding europium oxide with neutrons from the reactor. But as europium-152m decays to samarium-152 with a half-life of 9.3 hours (the time for half the nuclei in a sample to decay), they had only a few hours to use the europium-152m before so much of it had changed to samarium that measurements were no longer possible.

The second major problem the experimenters faced was to ensure that they detected only gamma-rays emitted in the direction in which the spinning samarium nucleus had been travelling. This, in turn, implied detecting gamma-rays that had lost no energy in making the nucleus change direction. To do this, the researchers used 'reflectors' made from samarium oxide. A samarium nucleus in the reflector would readily absorb any gamma-ray of exactly the right energy to set the nucleus spinning again. Then the nucleus would almost instantly emit a gamma-ray, so as to return to its normal, spinless state, in effect 'reflecting' the gamma-ray. Gamma-rays with the wrong energy would simply be transmitted through the samarium reflector.

A third essential part of the experiment was to measure the spin of the emitted gamma-rays, but this was relatively straightforward. The team could make the gamma-rays travel through magnetised iron. The ability of the gamma-rays to penetrate the magnetised iron would depend on the direction of their spins relative to the direction of the magnetic field. So by reversing the direction of the field, and then measuring which direction yielded the most gamma-rays, they could detect the polarisation – or 'handedness' – of the gamma-rays.

Goldhaber, Grodzins and Sunyar made nine separate measurements, some differing in detail from the others, but the message carried by the gamma-rays came over loud and clear. The gamma-rays were left-handed, and this meant that the neutrinos were left-handed.

The massive neutrino

> ... the discovery of non-vanishing neutrino masses would shed light on the theory beyond the standard model; thus it is one of the important issues experimentalists have to address.[26]
>
> *Jean-Luc Vuilleumier, 1986.*

Jeremy Bernstein, theoretical physicist and gifted writer, has described the events of 1956–7 as the 'Glorious Revolution'. The upheaval began with Yang and Lee's proposal that the weak force does not respect left–right symmetry, and it led to a new theory of massless left-handed neutrinos (and right-handed antineutrinos). In this theory the neutrino emitted in beta-decay has no mass. But what is the experimental evidence? Certainly it shows that the neutrino's mass must be very small, but is it zero?

When Pauli first tentatively proposed the existence of the neutrino in 1930, it was already clear that its mass could not be very large, as the electrons emitted in beta-decay occasionally take away most of the energy released by the nuclear transmutation. In his famous letter to the 'radioactive ladies and gentlemen', introducing the idea of neutrinos (which he referred to as *neutronen* or 'neutrons'), Pauli commented:

The mass of the *neutronen* should be of the same order of magnitude as the electron mass and in any case not larger than 0.01 times the proton mass.[27]

Three years later, Francis Perrin went farther. In making comparisons with experiment, he found that his calculations implied that the neutrino should have an intrinsic mass of zero. And then in his historic paper, 'Attempt at a theory of beta-rays', Fermi devoted section seven to 'The mass of the neutrino'. Here Fermi says:

The form of the continuous beta spectrum is determined by the transition probability. We shall first discuss how this form depends on the rest-mass μ of the neutrino in order to determine this constant from a comparison with the empirical curves ... The dependence ... is most obvious in the neighbourhood of the endpoints of the distribution curve ...

In Fig. 1 the end of the distribution curve is shown for $\mu = 0$ and for a small and for a large value of μ. The closest resemblance to the empirical curves is shown by the theoretical curve for $\mu = 0$.

We come therefore to the conclusion that the rest-mass of the neutrino is either zero or in any case very small with respect to the mass of the electron.[28]

He notes in addition that Perrin had come to the same conclusion.

Over the years, experiments investigated more and more precisely the endpoints of these beta-decay distributions and the results pushed the upper limit for the neutrino's mass inexorably downwards, to a tiny fraction of the electron's mass. But the experiments always yielded only a maximum possible value for the mass; they did not provide a lower limit and so could not resolve the question as to whether the neutrino has any mass at all.

Then in 1980, some intriguing news began to leak out of the Institute of Theoretical and Experimental Physics in Moscow (ITEP). The rumour was that a team there had managed to measure a very small, but nonzero mass for the neutrino. The timing was perfect, for the world of theoretical physics was poised to welcome such a discovery with open arms.

The previous few years had witnessed the steady rise of 'electroweak' theory – a single theory that describes the action of two of nature's fundamental forces, the electromagnetic force and the weak force. In one of the most significant developments since Fermi's theory of beta-decay, Abdus Salam, Sheldon Glashow and Steven Weinberg had independently pieced together a beautiful, self-consistent picture of the weak and electromagnetic interactions of all kinds of subatomic particle. And by 1980, experiments had time after time confirmed its validity.

Following this great breakthrough, the next logical step for theorists was to attempt to develop a larger, 'grand unified theory' that would bring the strong nuclear force under the same mathematical umbrella as the electromagnetic and weak forces. And this is where the interest in the neutrino's mass arose. In the electroweak theory there is no need for the neutrino to be anything more than the massless, left-handed object of the 'two component' theory. But grand unified theories could readily incorporate, and sometimes even demand, neutrinos with small masses. The mass of the neutrino began to assume enormous importance as a test of both the electroweak theory and the new grand unified theories.

The news from Moscow in the spring of 1980 was therefore tremendously exciting: Valentin Lyubimov and his colleagues claimed to have found that the mass of the neutrino must lie between 14 and 46 electronvolts (eV) – very much smaller than the electron's mass of 511 thousand electronvolts (keV), but nevertheless nonzero. There was an immediate flurry of activity among theorists across the world, as they worked out the implications of the results not only for the grand unified theories, but also for cosmology and astrophysics. The presence of neutrinos with even such a small mass could alter dramatically the balance of matter on a cosmic scale (see Chapter 7). But in the end, all this work was overshadowed by one question that for a long time failed to slip silently away: were the results from Moscow correct?

Weighing the neutrino

To be widely accepted a result of this importance coming from so delicate a measurement has to be confirmed by at least one other experiment.[29]

Jean-Luc Vuilleumier, 1986.

Measuring the mass of the neutrino is a very difficult experiment, and in some ways a thankless task. If the neutrino's mass is indeed exactly zero, you will never be able to prove this experimentally; the best you can do is to show that the mass is less than a certain amount, and that is not the kind of reward for years of hard work that appeals to many scientists. However, it is important for our understanding of neutrinos that we should try to find out as much as we can about them. And once Lyubimov's team appeared to have found a nonzero mass it was even more important that others should check their work.

The most direct way to measure the neutrino's mass is to follow Perrin and Fermi and to study the energy spectrum of the electrons emitted in beta-decay – in particular to look closely at the 'endpoint' where the electron's energy is at its maximum. As Fermi pointed out, the precise shape the spectrum takes as the number of electrons goes to zero depends critically on the neutrino's mass. But extraordinary difficulties face the experimenter wishing to measure this shape.

To begin with, very few electrons emerge with an energy close to the maximum, but the apparatus is swamped with unwanted, lower-energy electrons: the counting rate near the endpoint can be only a billionth or so of the total rate! Secondly, you need to measure the energy of the electrons with great accuracy, as this energy holds the key to the neutrino's mass; this implies knowing precisely what happens to electrons of a given energy as they pass through the apparatus. Thirdly, the source of the beta-electrons must be as thin as possible, as the particles lose energy in escaping through the material; but it must also give as high a rate as possible to compensate for the fact that you are working close to the end of the spectrum. And you must know what happens to the atoms or molecules that contain the beta-emitting nuclei. Can energy be lost in 'exciting' atomic electrons to a variety of energetic states, each giving rise to a slightly different beta-spectrum with a slightly different endpoint energy?

Another consideration is the choice of beta-emitter to study. The proportion of emitted electrons that have energies close to the endpoint is greater, the smaller the endpoint energy. So physicists interested in measuring the neutrino's mass have generally studied the decay of tritium, a heavy form of hydrogen, in which two neutrons accompany the single proton in the nucleus. Tritium decays by beta-emission with a half-life of 12.3 years. But of most interest for 'weighing' neutrinos is the energy at the far end of the spectrum, which is unusually low, with a value around 18.6 keV.

A key feature of experiments to measure the neutrino's mass is the spectrometer – the instrument used to measure the energies of the emitted electrons. In a simple magnetic spectrometer, a magnetic field bends the paths of electrons, spreading them out according to their energy, rather as a prism spreads out a beam of white light (see Chapter 2). A small aperture allows only those particles within a narrow range of energies to reach a detector, located at a point where the magnetic field brings particles of the same energy back to a focus. The inherent problem here is that a better

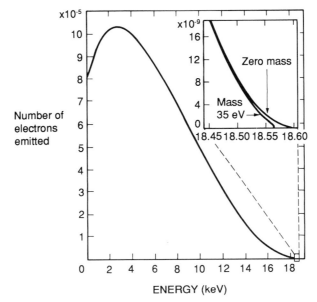

The energy spectrum of electrons emitted in the beta-decay of tritium. The insert shows the high-energy end of the spectrum magnified 2500 times to reveal the tiny effect expected if the neutrino emitted in the decay has a small mass (W. Kündig, University of Zürich.)

measurement of energy is possible only at the expense of intensity – that is, through reducing the size of the aperture, and through making the dimensions of the source smaller to reduce the initial spread of particles.

During the 1960s, Karl-Erik Bergkvist, at the University of Stockholm, developed an improved spectrometer that made it possible to work with much higher intensities than before. He introduced electric fields into the design, in particular to allow him to use a source with a large area, instead of a 'point' source. His technique was to use a varying electric field across the source. This altered the energy of the electrons in such a way that particles emerging across the source with the same energy were all focussed by the magnetic field to the same place. The rectangular source mapped onto a narrow line at the focus of the spectrometer, where the detector was located.

Previous measurements had shown that the mass of the neutrino must be less than 400–500 eV. With his extremely careful work, Bergkvist made an improvement by an order of magnitude. In 1972, he published his results, showing that the neutrino's mass is less than 60 eV, a remarkable achievement.

Then in 1975 E.F.Tretyakov at ITEP in Moscow proposed a more elaborate spectrometer, in which the electrons are dispersed and then refocussed four times in succession, passing through a sequence of apertures to scrape off unwanted electrons along the way. One beauty of this design is that it creates a relatively large distance – 2 metres or so – between the radioactive source and the electron detector. This helps to reduce the background of unwanted particles in the detector, which is an important factor when dealing with the small number of electrons near the end of the beta spectrum.

In March 1980, five years of work at ITEP bore fruit when Lyubimov, Tretyakov and their colleagues prepared to publish their results: they had found indications that the neutrino's mass lies between 14 and 46 eV. Physicists around the world were intrigued – and the debates began. Were there flaws in the experimental technique, or in the way that Lyubimov and his colleagues analysed their data? The consequences of a neutrino mass were, in any case, far too important for the result to go unchallenged. It had to be supported or refuted by one or more experiments.

Walter Kündig and colleagues at the University of Zürich were the first to publish new results, conflicting with those from Moscow. This team has built a spectrometer similar to Tretyakov's original design, but with some modifications. In particular, the instrument slows down the electrons in an electric field as they emerge from the source, so that they have around 10 per cent of their original energy. This means that for the same magnetic field, the spectrometer has greater spreading power and so picks off a narrower range of electron energies. To change the central energy selected by the spectrometer, the team simply changes the decelerating electric field, rather than the magnetic field, and this has the advantage of keeping the larger part of the instrument stable.

Another difference between the work at Zürich and that at Moscow lies in the source. Kündig and his colleagues have used tritium implanted in carbon that is evaporated onto a backing of aluminium foil. To create as large a source as possible, they use ten rings supported over a distance of several centimetres.

In May 1986, Kündig's team was ready to publish:

The [neutrino] mass determined is consistent with zero with an upper limit of 18 eV[30]

in a paper that ended:

In conclusion we find no indication of a non-zero mass . . . which is in strong contradiction to the results [from Moscow]. We see no possible source of error in our experiment large enough to account for this discrepancy.[31]

Meanwhile the team in Moscow had been busy improving their apparatus, modifying it so that they too could keep the magnetic field constant while using an electric field to select electrons of different energies. And in November 1986 they sent their new results to *Physical Review Letters*:

The combined analysis of both these data and the data of the previous cycle gives the neutrino mass 30.3 [+2, −8] eV.[32]

The discrepancy between the results from Moscow and those from Zürich remained.

However another paper had arrived at the offices of *Physical Review Letters* five days before the one from Moscow. This reported the results of a third team working with a similar spectrometer – a team at the Los Alamos National Laboratory in New Mexico, led by Hamish Robertson and Tom Bowles. Their main advantage was to have used a source of tritium gas, so they claimed:

The final-state effects in molecular tritium are accurately known and the data thus yield an essentially model-independent upper limit of 27 eV on the [neutrino] mass at the 95% confidence level.[33]

The tritium molecule is, relatively speaking, a simple system, consisting of two atoms, each with only one electron. With the aid of a reasonably powerful computer, it is possible to calculate exactly which energy states the molecule goes to when one of the tritium nuclei undergoes beta-decay. Moreover, the problems in knowing how much energy the electrons lose in escaping through the source material are much less severe with gas, which is much less dense than a solid. The price Bowles and Robertson and their colleagues have to pay is that the low density means a low decay rate, and that means taking longer to collect a sufficient amount of data.

At this stage, the results from Los Alamos did not completely resolve the difference between Moscow and Zürich. It was possible to argue, for example, that the data from both Moscow and Los Alamos would allow for the neutrino's mass to be only a little less than 27 eV, in contradiction to the result from Zürich. To resolve the issue required still more data, and by this time several teams around the world had begun to take up the challenge of measuring the neutrino's mass, using a variety of spectrometer designs and tritium sources. Their main aim – to develop apparatus that would reveal a neutrino mass down to around 10 eV. Good evidence that the mass must be less than this would lay the claims of Lyubimov's team to rest, even if it did not resolve the question of whether the neutrino has any mass at all.

By the summer of 1990, the evidence had begun to weigh heavily against Lyubimov's results. In June, at the *Neutrino '90* conference held in Geneva, the teams from Zürich and Los Alamos both presented new results. The physicists at Zürich had been working with a new source consisting of a layer of tritiated hydrocarbon only a molecule thick, bound to a smooth surface. With the new data, the upper limit for the neutrino's mass had fallen to 15.4 eV, with a confidence level of 95 per cent. The team from Los Alamos, meanwhile, claimed an upper limit as low 9.4 eV.

At the same time, further evidence in support of these new results was being accumulated in Japan. Here, a team based at the Institute for Nuclear Study at the University of Tokyo, has built a spectrometer of the kind Bergkvist used in his pioneering experiment. The Japanese group's main strength lies in using the same material as a source both when measuring the endpoint of the tritium spectrum and when measuring the response of the spectrometer to electrons of a known energy. They use the cadmium salt of an acid, so that to produce the tritium source they replace some of the normal hydrogen in the acid with tritium, while for the reference source they replace the cadmium with radioactive cadmium. The sources are remarkably thin, consisting of only two layers of molecules.

In February 1991, the journal *Physics Letters B* published the latest results from H. Kawakami and his colleagues. After carefully analysing 150000 decays, they had come to the conclusion that the neutrino's mass must be less than 13 eV, with a 95 per cent confidence level.

The results from Los Alamos, Tokyo and Zürich are pushing well below

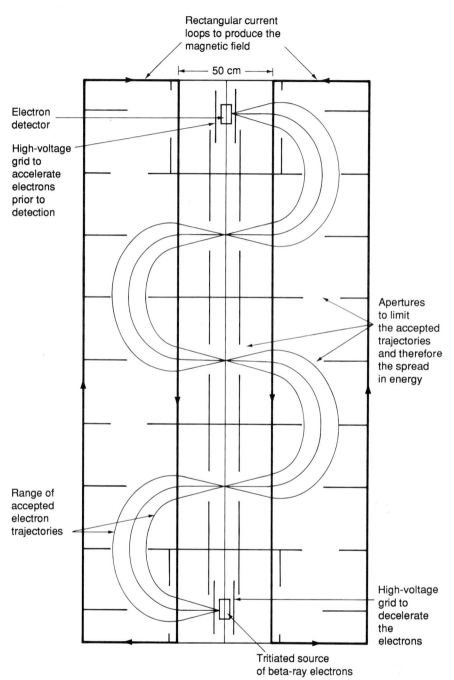

Rectangular current
loops to produce the
magnetic field

|← 50 cm →|

Electron
detector

High-voltage
grid to
accelerate
electrons
prior to
detection

Apertures
to limit
the accepted
trajectories
and therefore
the spread
in energy

Range of
accepted
electron
trajectories

High-voltage
grid to
decelerate
the
electrons

Tritiated source
of beta-ray electrons

The spectrometer used by the team at the University of Zürich to measure the mass of the neutrino emitted in the beta-decay of tritium. The diagram shows a cross-section through the cylindrical structure of the spectrometer. It indicates how this design, due to Tretyakov, resembles a series of adjacent spectrometers. As the electrons travel through alternating regions of magnetic field, they are spread out and refocussed four times and pass through several apertures before finally reaching the detector, 2.65 metres from the tritium source. (University of Zürich.)

the lower limit of 22 eV that Lyubimov's group published in 1987, and most physicists are coming to the conclusion that this result is wrong. Several experts, including the pioneer Bergkvist, have pointed out questionable procedures in the method that the team from Moscow uses to analyse its data. It now seems certain that the neutrino's mass lies below 15 eV. The only question that remains is how much lower?

The new generation of experiments should eventually set a limit as low as a few electronvolts. But unfortunately, exploring tritium beta-decay for a neutrino mass below that level becomes nigh-on impossible, because the effects on the spectrum would be so small. An experiment that took 2 years,

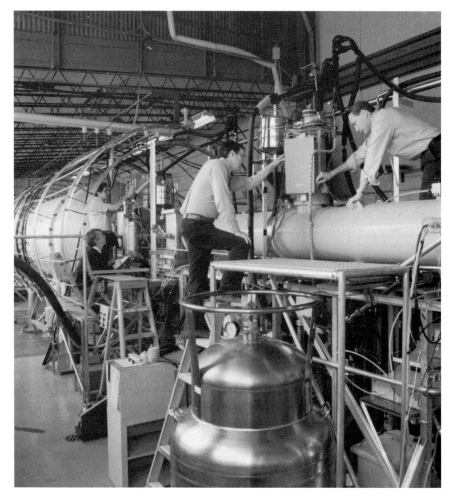

Tom Bowles (centre) and colleagues at Los Alamos check their apparatus to measure the energy spectrum of electrons emitted in tritium beta-decay. The electrons are guided through a long superconducting magnet, in the foreground, before passing into the spectrometer beyond, which is surrounded by a 'cage' of wires to cancel the Earth's magnetic field. (Los Alamos National Laboratory.)

say, to reach an upper limit to the mass of 10 eV, would take 2 million years to reach 1 eV! However, another type of experiment, based paradoxically on *not* observing neutrinos, is also helping to cast light on the neutrino's mass – and on the confusing conundrum of exactly what kind of neutrino the neutrino is.

Whose neutrino is it?

> In the late fifties and in the sixties the opinion was frequently expressed that neutrinos à la Majorana, although beautiful and interesting objects, are not realised in nature . . . [Since then] the question raised by Majorana has become more and more important and nowadays is, in fact, the central problem in neutrino physics.[34]
>
> *Bruno Pontecorvo, 1982.*

The charged particle emitted in the beta-decay of tritium is an electron. So, convention tells us, the neutral particle that also emerges is an *anti*neutrino. The convention at work here is a principle known as 'lepton conservation'. The electron and neutrino belong to a class of particles called *leptons* – particles that do not feel the strong force at work in the nucleus. The principle of lepton conservation, which is based on experience, says that in any interaction the total of leptons minus antileptons remains constant. Before a tritium nucleus decays, the only leptons around are the atomic electrons. But after the decay, an additional lepton – an electron – has been created. To balance the books, therefore, the other particle created must be an *anti*lepton – that is, an antineutrino.

Electrons and antielectrons (positrons) are clearly different particles, with opposite values of electric charge. It is not so obvious, however, that a similar distinction occurs with electrically neutral particles. The neutron, it turns out, does have a distinct antiparticle, the antineutron. But as we now know, a neutron is not a truly elementary particle; rather it is a conglomeration of electrically charged quarks. So, what about the neutrino?

This question was first raised in 1937 by Fermi's brilliant but unusual colleague at the University of Rome, Ettore Majorana. Bruno Pontecorvo, who became a student under Fermi in 1931, has recalled Fermi's admiration for Majorana and quotes him as saying:

Once a physical question has been posed, no man in the world is capable of answering it better and faster than Majorana.[35]

Born in 1906, Majorana had become known for his deep insight by the time he was 25, but according to Pontecorvo he was a pessimist who was permanently unhappy. From around 1934 he became more and more reclusive, and in 1938 disappeared for ever, having possibly committed suicide. Majorana did not bother to publish much of his work, but in the year before his disappearance, he decided to compete for a university chair, and wrote up a paper to improve his chances. Fifty years later this paper has now become famous, as it poses what Pontecorvo calls 'the central problem in neutrino physics': is the neutrino identical to its own antiparticle?

In the summary of his paper, Majorana writes:

The possibility is demonstrated of reaching a full formal symmetrization of the quantum theory of the electron and the positron using a new quantization process. This is modifying somewhat the meaning of the Dirac equations in the sense that these are no more reasons . . . to presume the existence of "antiparticles" . . . for new types of particles, especially neutral ones.[36]

In essence, Majorana's paper showed how his new method would leave the relationship between the electron and the positron intact, while simplifying the situation for neutral particles. While Dirac's theory led to four possible states – particle and antiparticle, each right-handed or left-handed – Majorana's theory led to only two states, right-handed or left-handed versions of a single particle. (As Giulio Racah soon pointed out, Majorana's symmetrisation of Dirac's theory in fact applied *only* to the neutrino, and not to the electron and positron as Majorana originally thought.)

Majorana commented in his paper that 'it is perhaps not possible now to ask to the experiment a choice'[37] between the theories, but as Pontecorvo has pointed out the idea of detecting neutrinos was not 'a decent argument of conversation' at the time. Only with the development of nuclear reactors with their huge emission of neutrinos did such tests become a possibility.

In 1946, while working at the Chalk River Laboratory in Ontario, it occurred to Pontecorvo that 'the appearance of powerful nuclear reactors made free neutrino detecting a perfectly decent occupation'.[38] He decided that the best technique would be to look for a form of inverse beta-decay that produced a radioactive nucleus; the number of neutrino reactions could then be measured by the induced radioactivity. And he found that the best reaction to go for would be one in which chlorine absorbs a neutrino and

Ettore Majorana, who proposed that the neutrino and the antineutrino might be indistinguishable, together with his sisters at Abbazia in Yugoslavia in 1932. (AIP Niels Bohr Library, gift of Erasmo Recami.)

becomes a radioactive form of argon, while at the same time emitting an electron.

Now, by the convention of 'lepton conservation', this reaction will occur only with neutrinos; one lepton (the neutrino) in effect turns into another (the electron). By the same convention a nuclear reactor emits *anti*neutrinos along with electrons from beta-decays. But in 1946, it was far from certain whether neutrinos and antineutrinos were indeed different. When Pontecorvo told Pauli of his idea,

> He liked very much the general idea and remarked that it was not clear whether 'reactor neutrinos' should definitely be effective in producing the reaction [with chlorine], but he thought that they probably would . . .[39]

In 1949, in the US, Luis Alvarez at the Lawrence Berkeley Laboratory in California independently made a detailed study of the same reaction. But in the end it fell to neither Pontecorvo nor Alvarez to put the concept into practice; instead, a chemist at the Brookhaven Laboratory took up the idea.

In 1955 Raymond Davis installed a detector containing four tonnes of carbon tetrachloride (dry-cleaning fluid) close to the reactor at Savannah River, at the suggestion of Reines who was working there on the famous experiment that discovered the neutrino. Chapter 6 describes in detail how Davis later perfected the remarkable techniques involved in this experiment to detect neutrinos from the Sun. Back in 1955, however, he found no evidence that antineutrinos from the reactor were absorbed in the chlorine detector. It seemed as though the particle emitted with an electron in beta-decay could not induce the emission of an electron in the inverse process. Neutrinos and antineutrinos appeared to be different kinds of particle, in accordance with Dirac's theory rather than Majorana's theory.

However, the discovery of parity violation provided an alternative explanation for Davis's result. The neutral particles emitted in beta-decay are right-handed, while those required for the inverse beta-decay of chlorine must be left-handed, *irrespective of whether neutrinos and antineutrinos are distinct particles*. The chlorine experiment can only distinguish left- from right-handed particles; it cannot tell the difference between Dirac and Majorana neutrinos.

Interest in Majorana neutrinos waned after 1957, with the success of the 'two-component' theory of neutrinos. But with the emergence of the grand unified theories in the late 1970s, theorists began to dust off Majorana's old ideas again, predicting the existence of Majorana neutrinos with small masses. Was there any hope of experimental evidence that might reveal the true nature of the neutrino?

If the neutrino has no mass, we can never tell. The theories of Dirac and Majorana will always give the same results. But if the neutrino has a small mass and is indeed its own antiparticle, as Majorana suggested, then there is one experimental test that can reveal it: the observation of a process known as 'neutrinoless double beta-decay'.

Virtual neutrinos

> From all this one would think, naively, that the difference between Dirac and
> Majorana neutrinos would stand out experimentally like a sore thumb [but] to
> this day, we still do not know which kind of neutrino actually exists.[40]
>
> *Jeremy Bernstein, 1984.*

As long ago as 1935, Maria Goeppert-Mayer, working at Johns Hopkins
University in Baltimore, investigated the possibility of an unusual form of
beta-decay, in which two neutrons change to protons almost simultaneously.
This 'double beta-decay' can occur only in nuclei that contain even numbers
both of protons and of neutrons. Some of these 'even-even' nuclei cannot
decay by normal beta-decay because the resulting nucleus would be heavier
rather than lighter than the initial one. (The total mass of a nucleus depends
not only on the number of particles it contains, but also, in a more subtle way,
on the balance of forces between them.) But by changing *two* neutrons into
protons, these even-even nuclei can reach states that *are* lighter, so double
beta-decay should be allowed. The process should emit two electrons and
two antineutrinos. However, as Goeppert-Mayer calculated, it must be very
rare, because it involves the weak force twice, and the overall effect must be
much weaker than in normal beta-decays.

Four years after Goeppert-Mayer's work, another theorist in the US,
Wendell Furry, realised that a different form of double beta-decay is
possible, if the neutrino is a Majorana particle. Furry's proposal was that, if
the neutrino is its own antiparticle, then the neutrino emitted by one neutron
(as in beta-decay) could be reabsorbed by another neutron (as in inverse
beta-decay), to give a net result of two protons plus two electrons, but no
neutrinos. The neutrino would play a *virtual* role, participating in the
process only at the unobservable quantum level. What is more, Furry
showed that, because the neutrinos do not have to emerge as real particles,

In 1935, Maria Goeppert-Mayer,
seen here with Enrico Fermi,
investigated the possibility of
double beta-decay, and deduced
that the process must be very
rare. (AIP Niels Bohr Library.)

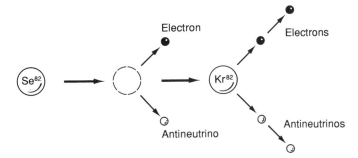

(a) Two-neutrino double beta-decay

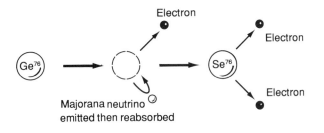

(b) Neutrinoless double beta-decay

(a) In double beta-decay, a nucleus with even numbers of both protons and neutrons can decay twice by beta-decay in quick succession, via a heavier intermediate state. This results in the emission of two electrons and two antineutrinos. In this example selenium-82 decays to krypton-82. (b) If the neutrino and antineutrino are indistinguishable, as Majorana proposed, the neutrino emitted in the first beta-decay will be reabsorbed almost instantly by the intermediate nucleus in a second process akin to beta-decay, which results in the emission of a second electron. The result, as illustrated here for the decay of germanium-76 to selenium-76, is neutrinoless double beta-decay.

the process should happen much more readily than the form of double beta-decay originally proposed.

However, the discovery of parity violation appeared to strike the same blow against Furry's argument, as it did against the results of Davis's experiment. If the object emitted by a neutron is right-handed, then it is never going to be absorbed by another neutron, so neutrinoless double beta-decay can never occur.

Or can it? What if the neutrino had some way of changing its handedness? It turns out that this is possible if the neutrino has some mass, even if it is very small. According to quantum theory, if the neutrino has mass then it must exist as a mixture of both possibilities – right-handed and left-handed. One way of thinking about this is to realise that a neutrino with mass can never travel at the speed of light, so, in principle, you can imagine overtaking it. Then if you look back at a left-handed neutrino, you will see it as a right-handed particle.

In this new picture, the neutrino emitted by a neutron has some small chance of being reabsorbed by another neutron, and neutrinoless double beta-decay *can* occur. And, more interestingly, the probability for it to occur depends on the mass of the neutrino – the heavier the neutrino, the more likely is neutrinoless double beta-decay to arise.

It seems a complicated argument of 'ifs' and 'buts'. However, the conclusions are sufficiently important to make worthwhile experiments that search for the rare process of neutrinoless double-beta decay. If an experiment discovers the process, then it will show both that neutrinos are indistinguishable from their antiparticles (they are Majorana particles), and that they have some mass, which will be revealed by the rate at which the process occurs. (There is, in fact, more than one way to overcome the 'handedness' problem, but the possibility of Majorana neutrinos is arguably the most exciting.) If the same experiment does not discover the process, at least it will say that neutrinoless double beta-decay cannot occur above a certain rate, and therefore if the neutrino is indeed a Majorana particle, then its mass must be less than a certain amount.

The way to observe double beta-decay is to detect the two electrons emitted in the process. But how do you tell whether or not two invisible neutrinos have been emitted at the same time? The answer lies, as in the original hypothesis of the neutrino's existence, in looking at the energy spectrum of the electrons. If double beta-decay produces two neutrinos, then a broad spectrum of electron energies will arise. But if there are no neutrinos, then in conserving energy and momentum the electrons can emerge only with a unique energy; the spectrum will be a spike rather than a broad continuum.

Double beta-decay *does* occur. Studies of rocks have shown that during

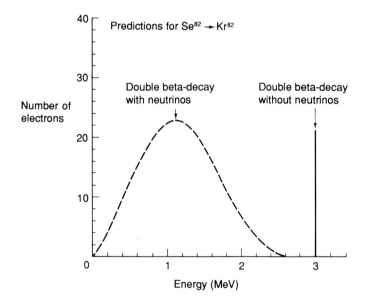

If double beta-decay results in the emission of two neutrinos, then these neutrinos will share the energy released with the two electrons, and the electron energy spectrum will be continuous, as it is for normal beta-decay. However, if no neutrinos are emitted, the two electrons will move away back-to-back, sharing the energy uniquely to produce a spectrum with a single peak.

the Earth's lifetime (about 4.5 billion years) the process has produced nuclei of inert gases. For example, xenon-130 (with 54 protons and 76 neutrons) has formed from the decay of tellurium-130 (52 protons and 78 neutrons). The decays are very rare. The half-life for tellurium-130 is estimated to be about 1.5×10^{21} years. In a small sample of solid tellurium-130, only a few decays would occur each year.

But how can you tell if this is double beta-decay with or without neutrinos? The only recourse in the case of geological samples is to theory, which indicates that the half-life is consistent with the decay being predominantly with neutrinos, rather than without. In a laboratory experiment, however, you can hope to measure the spectrum of the emitted electrons.

Michael Moe, and colleagues Steve Elliot and Alan Hahn at the University of California, Irvine, have done just that with selenium-82, which can decay to krypton-82 by double beta-decay. They have evaporated 14 grams of selenium onto a thin plastic (Mylar) foil and stretched it across the centre of a cylindrical chamber. The chamber is filled with gas, and subject to both an electric field (between the centre and the ends) and a parallel magnetic field. Any electrons emitted by the central plane move out towards the ends (the anodes), spiralling in the magnetic field as they go. The electrons leave ionised tracks in the gas, which drift in the electric field to induce a pattern of signals in wires at the ends of the chamber.

By correlating the time and position of these signals, Moe and his colleagues can work out exactly where in the foil the tracks originated. This gives the team a very powerful technique for searching for double beta-decay, for they can select the pairs of electrons that originate from the same place. Moreover, the curvature of the tracks in the magnetic field – translated into a spiral pattern of signals on the chamber ends – reveals the momentum of the electrons.

When they look at the energy spectrum for these electron pairs, the physicists do not find the sharp spike expected for neutrinoless double beta-decay. Instead, even with only a few hundred pairs, they clearly see the broad spectrum that indicates the presence of invisible neutrinos. Encouraged by their success, Moe and his colleagues are building an improved version of their detector, with 38 grams of selenium-82, which they plan to set up underground to reduced unwanted background signals. With this they will look more closely still for neutrinoless double beta-decay and measure the two-neutrino half-lives of other nuclei such as molybdenum-100.

The most popular way to search for neutrinoless double beta-decay, however, is to look for the decay of germanium-76 (32 protons and 44 neutrons) to selenium-76 (34 protons and 40 neutrons). Germanium is a semiconducting material made in a very pure form for the microelectronics industry. It is also a useful particle detector. And natural germanium contains 7.8 per cent germanium-76, so it automatically provides an all-in-one source and detector.

The decay of germanium-76 to selenium-76 would release an energy of

Michael Moe in front of his time projection chamber, which is surrounded by coils to produce the magnetic field. Moe and his colleagues at the University of California, Irvine, used this detector to measure for the first time the energy spectrum of the electrons emitted in double beta-decay. (M. Moe, University of California, Irvine.)

The energy spectrum for electrons emitted by selenium-82,
measured by Moe and his colleagues, shows the continuous
shape expected if neutrinos are
emitted together with the electrons in double beta-decay. (M.
Moe, University of California,
Irvine.)

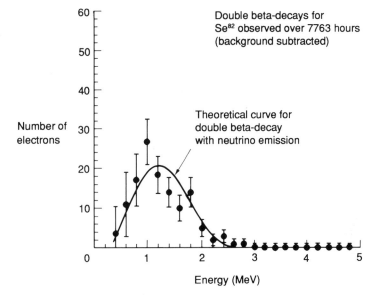

2.04 MeV, and in the case of a neutrinoless decay, this energy would go
entirely to the two electrons. But if neutrinos also emerge, the electrons will
have less energy to share between them. The technique with the germanium
detector, then, is to look for a spike at 2.04 MeV in the measured energy
spectrum.

The best results for this kind of measurement so far have come from an
experiment set up in the power house of the Oroville Dam in the Sierra
foothills in California. A team from the University of California at Santa
Barbara and the Lawrence Berkeley Laboratory has installed detectors here,
200 metres underground, to reduce the number of cosmic rays reaching their
germanium detectors. David Caldwell and his colleagues have eight large
(160 cubic centimetres) crystals of germanium mounted in a cryostat – a
vessel that keeps the germanium at the temperature of liquid nitrogen, to
reduce thermal effects. In addition, detectors made from sodium-iodide
crystals surround the germanium to reveal unwanted charged particles and
gamma-rays.

So far, Caldwell's team has found no evidence at all for neutrinoless
double beta-decay; the spectrum at 2.04 MeV shows no spike above the
usual statistical fluctuations. This allows the team to place an upper limit on
the lifetime of neutrinoless double beta-decay, which they say must be less
than 2.3×10^{24} years. This in turn allows them to place an upper limit on the
mass of the neutrino of about 1 eV. That is, *if* the neutrino is a Majorana
neutrino (if it is its own antiparticle), then if it has a mass it must be less than
1 eV. If the neutrino and antineutrino are distinct, as Dirac assumed, then
this experiment can make no comment.

Many experiments to search for neutrinoless double beta-decay are now

The team from the University of California, Santa Barbara, and the Lawrence Berkeley Laboratory, who have searched for neutrinoless double beta-decay in germanium-76 in a detector kept below ground in the power house at the Oroville Dam in California. (University of California, Lawrence Berkeley Laboratory.)

underway at various locations around the world, all beneath much greater shielding than the team has at the Oroville Dam. It seems, like much to do with neutrinos, a crazy game. Yet for many physicists it remains worth playing, because the stakes are so high. Indications of a neutrino mass would lead beyond electroweak theory, and perhaps open the way to the next important theoretical breakthrough. Meanwhile, although the true nature of the neutrino remains unresolved, physicists continue to refer to the left-handed versions as neutrinos, and the right-handed versions as antineutrinos. I will do the same throughout the remainder of this book, where the distinction is necessary. In the next chapter, however, we shall discover that even this distinction is not sufficient, for it turns out that there are more than one type of neutrino.

4

How many neutrinos?

... the sea-level penetrating [cosmic-ray] particles had this paradoxical behaviour. They seemed to be neither electrons nor protons. We ... resolved the paradox in our informal discussions by speaking of 'green' electrons and 'red' electrons – the green electrons being the penetrating type, and the red the absorbable type ...[1]

Carl Anderson, 1983.

DURING THE 22 YEARS between Fermi's 'tentative' theory of beta-decay, in which the neutrino was made respectable, and the experiment of Cowan and Reines, which proved that neutrinos exist by making them *do* something, a great deal happened in the realm of subatomic particles. Indeed, a new area of physics began to flourish, as the study of protons, neutrons and the like became less to do with the physics of the atomic nucleus and more to do with the nature of the particles themselves. During the 1930s and 40s, particle physics was born.

Many of the early discoveries came in experiments studying the cosmic radiation – that is, the rain of energetic particles that originates in outer space. High-energy protons and gamma-rays, in the main, arrive at the Earth's atmosphere after journeys from distant stars and probably even from other galaxies. When the new arrivals collide with atomic nuclei in the upper atmosphere they react violently, sometimes creating large showers of particles that cascade down towards the surface of the Earth. By ground level, many of these particles have been absorbed in further collisions in the atmosphere. But some remain, to pass through our bodies and into the Earth. Indeed, many neutrinos in the cosmic rain go right through the Earth and out the other side!

One type of particle that arrives in the cosmic rain at ground level is an 'alien' – a particle that seems unnecessary to matter here on Earth. In the mid-1930s, the list of necessary particles had seemed in good order. Atoms contain electrons, protons and neutrons; they interact by absorbing and emitting photons; and they transmute by emitting neutrinos in the company of electrons in beta-decay. Then in 1937, two American physicists, Carl

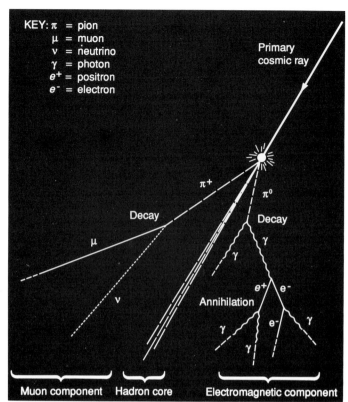

KEY: π = pion
     μ = muon
     ν = neutrino
     γ = photon
     e⁺ = positron
     e⁻ = electron

High-energy primary cosmic-rays from outer space – mainly protons – continually bombard the Earth's upper atmosphere, where they interact, producing showers of secondary cosmic ray particles, including pions and muons.

Carl Anderson, who together with Seth Neddermeyer discovered the muon, at the control panel of his cloud chamber on Pike's Peak in 1935. (AIP Niels Bohr Library.)

Anderson and Seth Neddermeyer, discovered the particle we now call the *muon*.

What is a muon? The best answer probably is that it is a 'heavy electron'. In short, the muon behaves very much like an electron, but is about 210 times as heavy. However, because the muon is heavier than the electron, it does not exist for ever. Instead, with a half-life of 2.2 microseconds, it decays into an electron and two neutrinos, *in the same way that a neutron decays into a proton, an electron and a neutrino.* I have used italics for these last words, as they hold the key to a major turning point in particle physics. But first some history is in order.

To begin with, soon after its discovery, the muon did appear to have a role. Late in 1934, a Japanese theorist had welded the ideas of Dirac, Heisenberg and Fermi together in a theory that accounted both for the force binding protons to neutrons and for the process of beta-decay. Hideki Yukawa was then a young lecturer at Osaka University, interested in developing a theory of the nuclear force, a difficult problem that caused him 'long days of suffering . . . the most difficult years of [his] life'.[2]

In the quantum theory of electromagnetism, the electromagnetic field is represented as a host of photons; two charged particles interact when one emits a photon and the other absorbs it. Yukawa had decided in 1932 to attempt a similar approach for the nuclear force field, arguing that:

. . . it seemed likely that the nuclear force was a third fundamental force, unrelated to gravitation or electromagnetism . . . [which] could also find expression as a field . . .

Then, if one visualises the force field as a game of 'catch' between protons and neutrons, the crux of the problem would be to find the nature of the 'ball' or particle.[3]

In his first attempt, Yukawa followed the work of Heisenberg, and used a field of electrons to supply the nuclear force between protons and neutrons. But this approach led to problems, concerning for example the spin of the electron. Then, in 1934, Yukawa decided to look no longer among the *known* particles for the particle of the nuclear force field:

The crucial point came . . . one night in October. The nuclear force is effective at extremely small distances, on the order of 0.02 trillionth of a centimetre. My new insight was the realization that this distance and the mass of the new particle I was seeking are inversely related to each other.[4]

Yukawa realised that he could make the range of the nuclear force correct if he allowed the ball in the game of 'catch' to be heavy – approximately 200 times heavier than the electron. Although such a particle was as yet undiscovered, he felt his suffering to be over, 'like a traveller who rests himself . . . at the top of a mountain slope.'[5]

The mass of Yukawa's particle was just the mass of the new cosmic-ray particle that Anderson and Neddermeyer discovered three years later – and which became known at first as the 'mesotron'. So when physicists in Europe and the US became aware of Yukawa's theory, they rapidly assumed that the mesotron was indeed the particle the theory required.

In 1935, Hideki Yukawa predicted the existence of a particle that is exchanged between protons and neutrons in the nucleus, as in a game of quantum 'catch'. The particle, now called the pion, was eventually discovered in 1947. (Associated Press.)

In Yukawa's theory, the field particle was also responsible for beta-decay, which he described with a two-step reaction, in contrast to Fermi's one-step process. According to Yukawa, a neutron turns into a proton as it emits a field particle; the field particle then turns into an electron and a neutrino. So, if the mesotron *was* Yukawa's field particle, it might be expected to decay into an electron and a neutrino. And it appeared for a while that this was certainly the case.

At the beginning of the 1940s, E.J.Williams and G.E.Roberts at the University College of Wales, Aberystwyth, published some photographs that seemed to show cosmic-ray mesotrons decaying into electrons. They recorded the tracks of particles passing through a cloud chamber – a device in which water droplets condense out on ionised tracks after supersaturated gas

In Fermi's theory of beta-decay, a neutron emits an electron and an antineutrino directly as it changes into a proton. Yukawa proposed, by contrast, that the neutron emits an intermediate particle (which he designated U^-) as it changes to a proton; this intermediate particle then decays almost immediately into an electron and an antineutrino, producing the same net result. The diagrams in which arrowed lines represent the particles show the participation of the U^- more clearly.

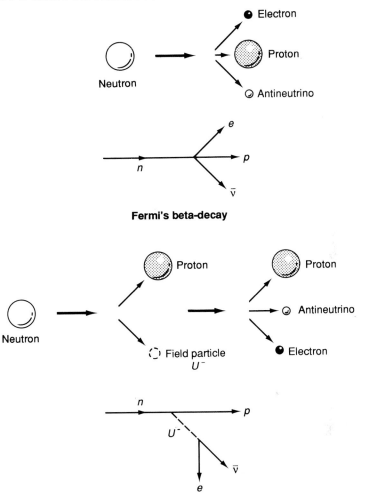

Fermi's beta-decay

Yukawa's beta-decay

inside the chamber is suddenly expanded. But it was difficult to capture the tracks of the lightly ionising electrons. More conclusive proof of the decay of the mesotron was still required.

In the years that followed, despite the Second World War, a number of physicists around the world turned to studying the mesotrons. Several experiments confirmed that the mesotron does indeed decay, with a half-life of a little more than 2 microseconds. What was perhaps the most significant result, however, came from an experiment performed by a group of Italians, who had been working in hiding after the Germany occupation.

While continuing their work after the Allied liberation of Rome in 1945, Marcello Conversi, Ettore Pancini and Oreste Piccione discovered that the cosmic-ray mesotron could not possibly be the field particle of Yukawa's theory. Briefly, negatively-charged versions of Yukawa's particle should have been captured readily by atomic nuclei in any matter the particles entered. Once attracted electrically to a positive nucleus, a negative mesotron should have been quickly swallowed into the nucleus by the strong nuclear forces. But the experiment showed that this effect did not occur. Whatever the mesotrons were, they were not influenced by strong nuclear forces. They could not be the particle Yukawa had invented to bind protons and neutrons together.

Shortly after the three Italians published their work, a team from Bristol University discovered another type of particle in the cosmic radiation, in an experiment high on a mountain top in the Pyrenees. Now the full story was revealed. The new particle turned out to behave exactly as expected for Yukawa's particle: furthermore, it evidently decayed into the more penetrating mesotron that had been studied so much at lower altitudes.

The heavy electron

> The fact that the large difference in the masses of the muon and the electron does not seem to induce other differences in their properties is one of the most fascinating in contemporary physics.[6]
>
> *Leon Lederman, 1963.*

Today, we call the particle that Cecil Powell and his team at Bristol found, the pion. The particle it produces when it decays, we call the muon. And we know, as Conversi and his colleagues first demonstrated, that they are very different. The pion, like the proton and the neutron, feels the strong force at work in the nucleus. The muon, like the electron, does not. But like the neutron, the muon does decay.

Precisely how much like the neutron became clear during the late 1940s, after Conversi, Pancini and Piccione had published their results. One of the physicists excited by their discovery was a former member of Fermi's group at Rome University, Bruno Pontecorvo, who was now working in Canada:

. . . as soon as I read [their] paper . . . I became fascinated by the particle we now call the muon. That was indeed an intriguing particle, 'ordered' by Yukawa, discovered by Anderson, and found by Conversi *et al* to be ill behaved to the point that it is nothing to do with the Yukawa particle! I found myself caught in an antidogmatic wind and I started to put lots of questions . . . who said that the muon must decay into an electron and a neutrino and not in an electron and two neutrinos, or into an electron and a photon? Is the charged particle emitted in the muon decay an electron?[7]

Pontecorvo had noticed immediately a close similarity between the rates at which electrons and muons are captured by nuclei. The process of electron capture is akin to beta-decay – a proton inside a nucleus catches a nearby electron and turns into a neutron, at the same time emitting a neutrino. So, in a paper sent to *The Physical Review* in June 1947, Pontecorvo proposed that when a proton captures a muon, it turns into a neutron and emits a neutrino.

The following year, Giampietro Puppi in Italy and Oskar Klein in

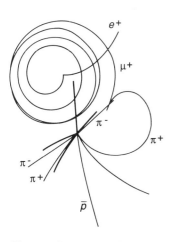

The complete process in which a positive pion (π^+) decays to a positive muon (μ^+), which in turn decays to a positron (e^+), is captured on this photograph of particle tracks taken in a streamer chamber. The pion has been produced in a spray of particles at the centre, in the annihilation of a proton and an antiproton. The tracks curl as the particles bend under the influence of a magnetic field, and the abrupt changes in direction reveal the decays, in which invisible neutrinos are also emitted. (G. Piragino, PS179 experiment at CERN.)

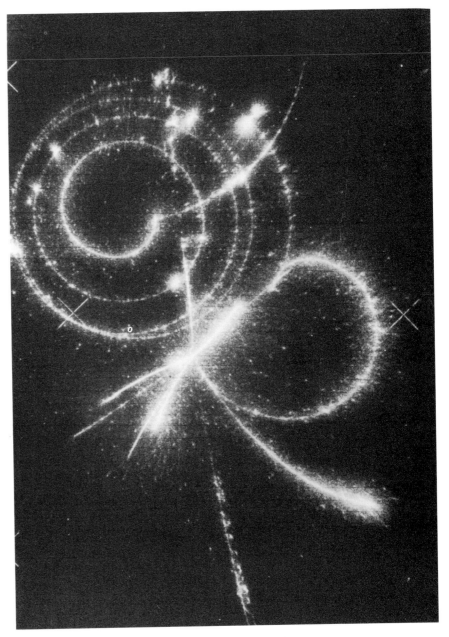

Stockholm working independently, also noticed how the capture and decay of muons appears to be closely related to beta-decay processes. They found that they could rewrite Fermi's theory for beta-decay in terms of the particles involved in the muon interactions. The most interesting point, however, concerned the 'coupling constant' – a constant that gives the strength of the interaction between the initial and final particles.

The coupling constant for the decay of the muon turned out to have almost the same value as the one calculated for beta-decay. This discovery implied that the same basic interaction is taking place – that the same force is responsible for the deaths of the neutron and the muon. And this, in turn, suggested the existence of a new force that works universally on different kinds of particle. This force became known as the 'weak force', as the phenomena it gives rise to occur much more slowly than reactions due, for example, to the electromagnetic force between particles. These slow interactions reflect the relative smallness of the 'coupling constants', and, in turn, the weakness of the force.

Meanwhile, Pontecorvo was 'hooked' on muons, and had taught himself as much as he could about cosmic-ray physics. He had teamed up with Ted Hinks at the start of a 'very friendly, unforgettable and stimulating experimental collaboration'.[8] In a short time, they had prepared an experiment to investigate muons and in particular to search for answers to the questions that so intrigued Pontecorvo.

They were working in reactor physics at the Chalk River Laboratory in Ontario, and had 'developed a sort of feeling of guilt in doing cosmic-ray research'.[9] Nonetheless, they found the answers to their questions. Yes, the charged particle emitted when the muon decays *is* an electron; no, the muon does not decay into an electron and a photon; no, the muon does not decay into an electron and a neutrino. In fact, the muon decays, like the neutron, into three particles, as Pontecorvo and Hincks observed through measuring the spectrum of the emitted electron, which is continuous just as in beta-decay. Other researchers found similar results. In particular, in the US, Jack Steinberger at Chicago University also found that the muon decays to three particles; but more of Steinberger later.

One neutrino, or two?

> [In 1956] Cowan and I proposed to go to an accelerator and test the identity [of the two neutrinos]. The reaction we got from Los Alamos was difficult to understand . . . 'You two fellows have had enough fun. Why don't you go back to work.'[10]
> *Fred Reines, 1982.*

Ten years after the experiments with Hincks, Bruno Pontecorvo was working in the Joint Institute for Nuclear Research at Dubna, south of Moscow, having left Canada for the Soviet Union in 1950. The beginning of 1959 saw him considering what research would be possible at the proton accelerator that the institute was proposing to build. One idea was to make neutrinos.

Bruno Pontecorvo, pictured here around 1948, made several contributions to ideas about neutrinos, from experiments on the decays of muons, to proposals to study neutrinos at nuclear reactors and at accelerators, to the suggestion that neutrinos might oscillate from one variety to another. (AIP Niels Bohr Library, Physics Today Collection.)

The first particle accelerators had come to life in the early 1930s. These machines give repeated bursts of energy to bunches of particles, usually electrons or protons. The result is an intense, controllable source of particles of far greater energy than those emitted by radioactive materials – indeed, more similar to cosmic rays of all but the highest energies.

While still in Canada, in 1946, Pontecorvo had realised that a nuclear reactor would make a reasonable source of neutrinos and he had begun to consider how to detect neutrinos. (This led him to propose using chlorine-37 as a neutrino detector, an experiment that Raymond Davis later performed, as described in Chapter 3; Cowan and Reines, who were apparently unaware of Pontecorvo's work, came independently several years later to the choice of a reactor as a neutrino source and used the more-recently discovered liquid scintillator in their detector.) With his move to Dubna, Pontecorvo had left the physics of nuclear reactors behind him. But by 1959, he was thinking about neutrinos again, this time made at an accelerator.

By the late 1950s the manufacture of pions at accelerators had become standard. The high-energy protons from the accelerator strike a metal 'target' and release large numbers of energetic pions, just as in the collisions of cosmic rays high in the atmosphere. The pions, in turn, decay to muons and neutrinos. Pontecorvo decided to make use of these neutrinos.

One experiment that he thought possible was to investigate the nature of the neutral particles emitted in the decays of muons and pions. His experiment with Hincks had shown that when a muon decays, two neutral particles accompany the electron that is produced. But are these two neutral particles really neutrinos? And if so are they both the same kind of neutrino? Moreover, the pion decays to a muon and a neutral particle. Is this neutral particle the same as the neutrino emitted in the beta-decay of a nucleus, or is it something different?

At the time, there was a particularly good reason why there might be more than one type of neutrino, and this concerned a crisis related to the decay of muons. Physicists by now believed that the muon likes to decay into an electron, a neutrino and an *anti*neutrino. They also knew that particles and antiparticles, such as electrons and positrons, like to annihilate – in other words, their quantum properties exactly cancel and their total mass converts into energy in the form of gamma-ray photons. So why, physicists asked themselves, do the neutrino and antineutrino produced in the decay of a muon not immediately annihilate and create a photon alongside the electron?

As Pontecorvo and Hincks had found, the muon apparently never decays into an electron and a photon. Indeed, several more sensitive experiments confirmed this. Yet, there seemed to be no reason why the muon should not decay in this way. On the contrary, explanations for a different problem concerning the weak interactions, suggested that such decays should occur thousands of times more frequently than was conceivably consistent with the experimental lack of observation.

However, there was one solution to the paradox. Suppose that the neutrino and antineutrino produced in the muon's decay are not of the same type. In this case they should not be able to annihilate, for their quantum properties will not cancel exactly. But how could the two kinds of neutrino differ?

In answering this question, physicists were guided by the observation that physical processes are governed by conservation laws. For example, all the processes we observe conserve energy and momentum overall; we observe none that disobey this rule. (Recall the consternation of Pauli when Bohr suggested that the problems of nuclear physics might be resolved if energy were not conserved.) So it seemed plausible that the muon does not decay to an electron and a photon because to do so would violate some fundamental conservation law of nature. The problem was to discover which property was being conserved, and the arguments went something as follows.

Suppose that the neutrino produced by the decaying muon carried on some property of the muon, a kind of 'muon-ness'. The antineutrino produced at the same time, being of a different type, would have zero 'muon-ness', but it could instead carry a related property of 'electron-ness'. This property for the antineutrino could cancel with the 'electron-ness' of the electron created in the decay – as the electron is a particle, while the antineutrino is an antiparticle. In this way, overall 'muon-ness' would be conserved through the production of the neutrino, while overall 'electron-ness', which was zero before the muon's decay, would remain at zero with the production of the antineutrino–electron pair.

Moreover, something similar would occur in nuclear beta-decay, where an antineutrino is born together with an electron. This antineutrino would

The fact that the neutrino and the antineutrino emitted when a muon decays do not annihilate to produce a photon can be understood if they are of different intrinsic types, related to the electron and the muon. It appears that in weak interactions, such as the decays illustrated here, muon-like particles (muons or muon-neutrinos) can decay only to other muon-like particles, and can be created only in the company of other muon-like *anti*particles; the same is true for electron-like particles (electrons and electron-neutrinos).

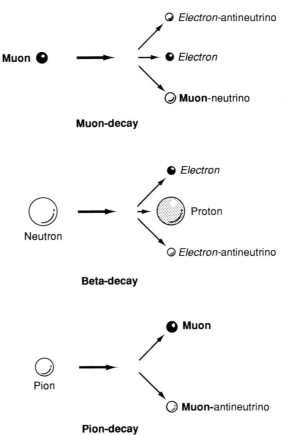

carry 'electron-ness'. On the other hand, the particle produced together with a muon when a pion decays would by similar arguments carry 'muon-ness'. Thus 'electron-neutrinos' and 'muon-neutrinos' would be two distinct types of particle.

To test for any difference between 'electron-neutrinos' and 'muon-neutrinos', Pontecorvo proposed using the neutrinos that are emitted along with muons in the decays of pions to bombard a detector similar to that used by Cowan and Reines. The key would lie with counting how many electrons were produced in 'inverse reactions' in which the neutrinos were absorbed. If the muon-neutrino was indeed different from the neutrino that accompanied the electron in beta-decay, then no inverse reaction would occur and no electrons would be detected. The accelerator that Pontecorvo was planning experiments around was never built, but across the world, in New York, a young American physicist had conceived of a similar idea.

Melvin Schwartz was a member of the physics department at Columbia University at a remarkable time. Like the Cavendish Laboratory in Cambridge under Rutherford, or the physics department at Rome under Fermi, Columbia University had become a veritable 'cauldron of ideas'. One

of the brightest stars in the galaxy of talent at Columbia at the time was
Tsung Dao ('T.D.') Lee, the young theoretical physicist who in 1957 had
shared the Nobel Prize with Chen-Ning Yang. They had shown how the
weak force does not respect mirror symmetry, the phenomenon known as
parity violation (see Chapter 3). Lee was now interested in how the weak
force behaves at energies higher than those involved in the decays of particles
such as muons and neutrons. And he would discuss his ideas with his
experimental colleagues – especially at coffee time!

Schwartz recalls:

One Tuesday afternoon in November, 1959, I happened to arrive late at coffee to find
a lively group around T.D. . . . with a conversation ensuing as to the best way of
measuring weak interactions at high energies. A large number of possible reactions
were on the board making use of all the hitherto known beam particles – electrons,
protons, neutrons. None of them seemed at all reasonable . . .

That night it came to me. It was incredibly simple. All one had to do was use
neutrinos . . .[11]

When it came to studying the weak force, neutrinos offered a specific
advantage over other types of particle. They have no electric charge, and
they do not feel the strong nuclear force; indeed, the only way that neutrinos
can interact with anything else is through the weak force. So, in experiments
with neutrinos, the effects of the weak force would not be obscured by
stronger electromagnetic or nuclear effects. And Schwartz believed he knew
how to make neutrinos.

Schwartz had hit upon the same basic idea as Pontecorvo, and he could see
how to use it to create a beam of neutrinos. Energetic protons striking a small
metal target produce pions that continue mainly in the same direction. As the
pions decay, they, in turn, produce neutrinos, which again tend to follow the
direction of the pions, although spreading out slightly. At the energies that
were typically available at the biggest accelerators in the late 1950s, the pions
would on average travel a distance of 50 m before decaying, while the
neutrinos produced would tend to continue within an angle of 10° to the
main direction. To avoid the neutrinos being too widely spread, Schwartz
proposed placing a 10 metre thick shielding wall about 10 metres from the
target where the pions were formed. About 10 per cent of the pions would
decay in this distance. The remainder, as well as the muons produced in the
decays, would be absorbed in the shielding wall. Only the neutrinos that had
been born from the decays of the pions would continue through the wall.

Schwartz then calculated how many protons he would need to make a
neutrino beam of sufficient intensity to produce a few reactions in a detector.
Here the high energy of the neutrinos relative to those produced in beta-
decay would help. As two Japanese theorists, Sin-itiro Tomonaga and
Hidehiko Tamaki, had shown back in 1937, the probability for a neutrino to
react increases rapidly with its energy. The cross-section rises a million-fold
from around 10^{-44} cm^2 – the value that Bethe and Peierls calculated (see
Chapter 3) for neutrinos from a reactor with energies of a few MeV – to

Tsung-Dao Lee, lecturing on the weak force at the European laboratory CERN, in 1965. While working with Yang on theoretical ideas on the high-energy behaviour of the weak force, Lee inspired Mel Schwartz, an experimental colleague at Columbia University, to propose a means of producing a beam of neutrinos to study weak interactions. (CERN.)

around 10^{-38} cm² for neutrinos with energies of a few hundred MeV, within the capabilities of the highest energy accelerators of the time.

Assuming that he could build a detector containing 10 tonnes of material, Schwartz reasoned that he would need an accelerator capable of delivering five million million (5×10^{12}) protons a second in order to produce one reaction an hour in the detector. This was far more protons than any accelerator at the time appeared capable of producing. However, people were discussing how to build new machines with higher intensities, so Schwartz published his ideas in March 1960, in *Physical Review Letters*. He had, naturally, discussed his ideas with T.D. Lee, and Lee together with his colleague Yang, had written a paper on what one could learn from experiments with a neutrino beam. This appeared in *Physical Review Letters* immediately following the paper by Schwartz. First on the list of possibilities was the suggestion of testing the identity of the neutrinos produced in pion decay and neutron decay.

The first high-energy neutrino experiment

> By the grace of the AEC, BNL, God, Green and Hayworth (alphabetical order), we should see neutrinos.[12]
>
> *Leon Lederman, 1960.*

Even if it did not appear to be immediately possible, the idea of a neutrino beam seemed too good to let go, at least as far as some of the physicists at Columbia University were concerned. In particular, Leon Lederman and Jack Steinberger (who had been Schwartz's thesis supervisor) refused to put the suggestion on the shelf. Both were well acquainted with Lee and were ready to accept his challenge to extend the studies of the weak force into a new domain.

Lederman and Steinberger had between them already performed many important experiments. In 1949, in his thesis experiment at Chicago University, where he worked under Fermi, Steinberger had shown that the muon decays to three particles, one charged and two neutral, just as Hincks and Pontecorvo discovered independently around the same time. Lederman, a graduate student at Columbia University from 1948–51, had in 1957 studied the symmetry properties of muon decays at the behest of Lee and Yang. This experiment had confirmed that weak decays violate spatial symmetry as Lee and Yang had proposed. And in the same year, Steinberger's group at Columbia University (including Schwartz) – again encouraged by Lee – found a related asymmetry in the weak decay of hyperons. (These are particles that are akin to protons and neutrons, but which are heavier and therefore short-lived.)

Together with Schwartz, Lederman and Steinberger kept working at the idea of setting up a neutrino beam. In particular, they had their sights set on the new accelerator that was being built nearby on Long Island at the Brookhaven National Laboratory, a place where they had already worked on the older accelerator, the Cosmotron. The new machine was designed to accelerate protons to 30 GeV (30 giga electronvolts or 30000 MeV) – ten times the energy that the Cosmotron reached – but would it produce enough protons to yield a useful number of neutrinos? Perhaps. 'We kept arguing that . . . maybe the Brookhaven accelerator was good enough,' recalls Lederman, 'and ultimately, we hit on a scheme for doing the research.'[13]

One clear requirement was for a large detector to stop as many neutrinos as possible, even if that meant only a few a day. But just how would you build a detector weighing 10 tonnes or so? Fortunately, not far away in New Jersey, at Princeton University, James Cronin and George Renniger were working with a new type of detector which offered a solution – the spark chamber.

A simple spark chamber consists of a series of parallel metal plates enclosed in a box filled with a suitable gas. When a charged particle passes through the chamber, it leaves behind a trail of ionised atoms. If a high electric field is now set up between pairs of plates, the gas will break down and spark where ionised. In this way, the particle's trail between the plates becomes revealed as a chain of sparks, which you can photograph.

In 1960, the principle of the spark chamber had only recently been demonstrated, first in Britain and then in Japan. Cronin and Renniger were in the process of testing one of the first chambers to be built in the US – a box of 19 plates of aluminium, 6 millimetres thick, each with a useful area of about 320 square centimetres. The total weight of the aluminium was in the region of 10 kilograms. Yet to Schwartz and his colleagues it was at once apparent that here was the ideal device for detecting neutrinos.

For one thing, a spark chamber has a memory. The ionised trail remains until you apply a high voltage to create the electric field. So you can use other nearby detectors to show whether anything had emerged from the chamber, and only then use the signals from these detectors to trigger the high-voltage supply and make the chamber spark. In this way, you need trigger the chamber and take photographs only when it seems likely that something interesting has happened – a useful option when working with neutrinos, which rarely interact.

Secondly, you could build a 10-tonne spark chamber without its being outrageously large and without spending a huge amount of money. The metal plates would provide the mass; it was simply a question of scaling up. 'Simply' is rather an overstatement. It was more a question of audacity. At a conference in Berkeley in September 1960, Cronin presented some of his first pictures of particle tracks in his 10-kilogram chamber, obtained only the month before; at the same conference, Lederman presented the first thoughts on a 10-tonne spark chamber, estimated cost around $70 000.

By this time the new accelerator at the Brookhaven Laboratory had begun to operate, and the likelihood that it would produce proton beams intense enough to create a useful number of neutrinos grew with each day. The neutrino group at Columbia set to work with a will, Steinberger and Lederman leading a team that finally consisted of graduate students Jean-Marc Gaillard, Dino Goulianos, and Nariman Mistry together with Gordon Danby, a physicist from the Brookhaven Laboratory, and of course Schwartz.

The scaled-up spark chamber was roughly ten times bigger in each

Mel Schwartz standing in front of the 10-tonne spark chamber used in the 'two-neutrino experiment'. Each of the ten modules contains 1 tonne of aluminium in the form of nine plates which are 2.5 centimetres thick and separated by a gas-filled gap of 1 centimetre. High voltage across the plates causes the gas to spark along the tracks of charged particles, which, in this time-lapse picture, are cosmic rays. (Brookhaven National Laboratory.)

dimension than Cronin's prototype. It contained 90 plates of aluminium, each 2.5 centimetres thick and 120 centimetres square. These were grouped together in ten 1-tonne modules, of nine plates each. In each module, plastic 'window frames' between the plates held them 1 centimetre apart, and helped to form the outer walls of the module so that it could be filled with neon gas. The ten modules were arranged in two rows of five, stacked one on top of the other.

Neutrinos themselves would not leave ionised trails in the modules, but any charged particles – that is, muons or electrons – created by neutrinos in interactions in the aluminium plates, would produce tracks. The aim was to trigger the high-voltage supply whenever such charged particles occurred,

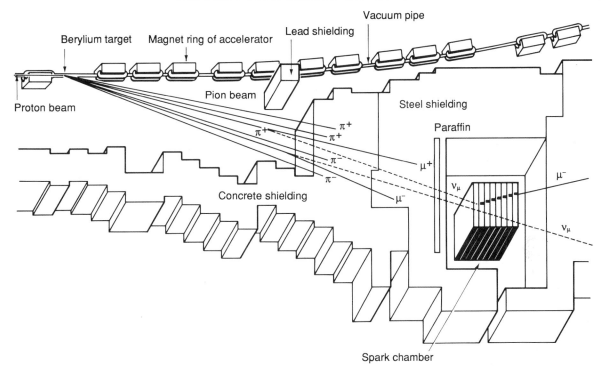

The principle of the first high-energy neutrino experiment at Brookhaven was to create the neutrinos in the decays of pions produced when protons in the accelerator struck a target of beryllium. Large amounts of steel shielding in a wall 13.5 metres thick absorbed both the muons produced and the remaining pions, allowing only the neutrinos to penetrate to the 10-tonne spark chamber.

so as to make the tracks visible as trails of sparks. The luminous trails could then be photographed by cameras viewing the sandwich of modules from the side.

To detect charged particles, the team installed slabs of plastic scintillator, between each module and at the back of the detector. The scintillator would emit light when charged particles passed through, and photomultiplier tubes attached to one end of the slabs would rapidly convert the flashes to electric signals, a few tens of nanoseconds (thousandths of a microsecond) after the charged particles had passed through the chamber. These signals could then trigger the high voltage on alternate plates in the spark-chamber modules, setting up the electric fields that would cause the sparks to form.

Building the spark-chamber modules was only one of the tasks the team faced. Like Cowan and Reines, they had to work out how to prevent particles other than neutrinos from rendering the experiment useless. One solution was to shield the detector with enough material to keep out high-energy particles, particularly muons. The pion beam that yielded the neutrinos was allowed to travel about 20 metres before it crashed into a wall of steel 13.5 metres thick. Some 10 per cent of the pions would decay before reaching the wall, which would stop the remaining 90 per cent of the beam as well as the penetrating muons produced in the decays. Only the neutrinos would emerge from the far side. The wall itself was built from armour-plate from the deck of an old battle ship. At this time the Nevis Laboratory at Columbia University was funded by the US Navy, and could acquire scrap metal cheaply.

But this wall alone was not enough. Particles would get round it and into the detector. So the researchers had to install more steel in a side wall that ran close to the electromagnets of the proton accelerator – much to the initial alarm of Kenneth Green who was in charge of the new machine. On the remaining sides, floor and roof, concrete provided still more shielding.

However, several hundred cosmic-ray particles could still enter the detector each second, despite the shielding, and these could pass through the slabs of scintillator used to trigger the spark chambers. To avoid triggering unnecessarily the team installed more slabs of scintillator above the detector and to the front and back. The signals from these scintillators were used to indicate unwanted particles and to 'veto' the triggering of the spark chambers. This reduced the numbers of triggers due to cosmic rays to about 80 per second. But the sides of the detector had to remain unobscured so that cameras could photograph the tracks.

A final trick to reduce the number of unwanted tracks called on the skills of the accelerator crew under the direction of Ken Green. Genuine tracks from neutrino interactions would occur only very shortly after protons had emerged from the accelerator to generate the neutrino beam. So the physicists designed their electronics to trigger the spark chambers only if the accelerator was actually delivering protons. And by reducing the time intervals over which the protons emerged to the shortest times possible, the maximum number of cosmic-ray 'accidental' triggers was avoided.

In the end, the detector accepted particles during 1.6 million pulses of protons from the accelerator, mainly on 25 'good' days, spread over a period of eight months or so. In this time, the detector was 'on' for a total of only 5.5 seconds, time enough for 440 cosmic-rays to trigger the spark chambers. But during the same time, nearly 10^{14} neutrinos passed through the detector, enough to produce, the team estimated, 25 reactions.

The result was 5000 photographs, more than half of which were blank, presumably triggered by the interactions of relatively slow neutrons. Other photographs the physicists recognised as being due to the expected 400 or so cosmic-rays, which came through gaps between the veto counters; and a large number of the rest appeared to be due to muons that had penetrated the shielding and missed the veto counters. But what did the remainder reveal? They certainly indicated that a few neutrinos had interacted in the detector, but in what way, and what had they produced?

In identifying particles through their tracks in the spark chambers, Lederman, Steinberger and their colleagues were greatly helped by the difference in mass between the muon and the electron. When an electron travels through a material, it loses energy rapidly, radiating as it comes under the influence of the electric fields of atomic nuclei. The electron comes quite

The long straight track of a muon is made visible as a series of short sparks between the aluminium plates, once the particle has crossed the 10-tonne spark chamber. The shorter track above is probably due to another particle originating from the same neutrino interaction that produced the muon. (Brookhaven National Laboratory.)

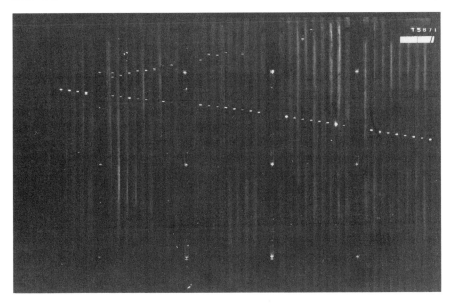

quickly to halt, staggering around as it scatters from the atoms it encounters. Because muons are much heavier they radiate nowhere nearly so easily. They travel far greater distances than electrons of the same energy before stopping, and they leave straight tracks as they travel on unswervingly.

In the final analysis, the team at Columbia were left with 29 photographs showing a long straight track, typical of a muon, appearing to come from nowhere. This implied that the 'muon-neutrinos', born with muons in the decays of pions, could produce muons. But could they also produce electrons? Electrons would leave more erratic, staggered tracks, as the researchers confirmed by testing two of their spark-chamber modules in a beam of electrons at the Cosmotron, the 3-GeV accelerator, which was still operating at Brookhaven. However, the pictures gathered in the chamber as a whole failed to reveal more than a few examples of tracks that might have been due to electrons. There was certainly nothing like the 29 or so pictures of electron tracks that might have been expected were this neutrino beam equally as good at creating electrons as muons. 'The most plausible explanation', the researchers wrote when they published their results, 'is . . . that there are at least two types of neutrino' – those associated with electrons and those associated with muons.

And then there were three!

> I was following an old idea in science: 'If you can't understand a phenomenon, look for more examples of that phenomenon.'[15]
>
> *Martin Perl, 1980.*

The discovery that there are two separate types of neutrino, carrying 'electron-ness' and 'muon-ness', at least brought some symmetry into the picture that particle physicists were slowly building of the fundamental nature of matter. Now there seemed to be a distinct 'family' of particles – the electron, the muon and the two neutrinos – which can partake in electromagnetic processes and in weak interactions, such as beta-decay, but which are oblivious to the strong force at work in the atomic nucleus. These particles became known collectively as *leptons*, from the Greek for 'slender', as they were all significantly lighter than the proton and most other particles.

However, the discovery of the muon-neutrino did little to resolve the underlying question: why does nature repeat itself? Why does a heavier charged lepton – the muon – exist at all, when in every other respect it behaves just like an electron? And why should there be a second type of neutrino, perhaps also heavier than the electron-neutrino?

One person who was interested in these questions was Martin Perl, a physicist at the Stanford Linear Accelerator Center (SLAC) in California. SLAC boasts the world's longest linear accelerator, or linac, which was for many years also the most powerful electron accelerator. Early in 1972, the linac began a new task – feeding a machine called SPEAR, for 'Stanford Positron Electron Asymmetric Rings' (although in the final design it consisted of only one ring).

SPEAR is basically a ring of electromagnets threaded by a beam pipe, which is designed to store hundreds of billions of particles fed to it by the linac. The linac initially supplies electrons, which travel clockwise around SPEAR. But once there are sufficient electrons circulating, the linac begins to provide positrons (antielectrons), which travel round SPEAR in the opposite direction. The positrons and electrons annihilate when they meet, but only when the two beams of particles collide at locations on opposite sides of the ring.

When SPEAR started up in 1972, a team from SLAC and the Lawrence Berkeley Laboratory was ready and waiting. They had built a complex detector, called Mark I, which encircled the beam pipe at one of the collision regions. The aim was to detect the particles that rematerialised from the energy released each time an electron and a positron annihilated. Concentric layers of cylindrical spark chambers were there to reveal the tracks of charged particles; beyond these the aluminium coil of a solenoid provided the magnetic field to bend the tracks of the particles; then there was a layer of plastic scintillator and lead where gamma-rays and electrons would deposit most of their energy; finally, there were layers of iron and concrete, to stop all particles except muons and neutrinos, and an outer layer of spark chambers to detect muons.

The annihilation energy from collisions at an electron-positron collider such as SPEAR can rematerialise as almost anything, provided various conservation rules are not violated. For example, the total electric charge of the new particles must remain the same as before (zero); the number of particles and antiparticles must be equal; and the total mass-energy of the products must equal that of the colliding electron and positron. Sometimes the energy produces simply a new electron and a new positron, sometimes it

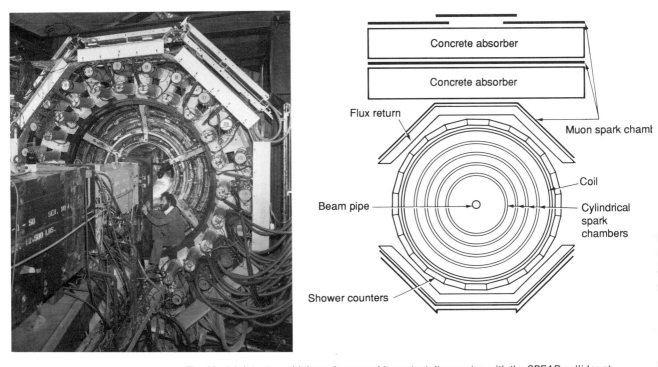

The Mark I detector, which made several important discoveries with the SPEAR collider at the Stanford Linear Accelerator Center in 1974–5. The diagram shows the different concentric layers that made up this basically cylindrical detector, built to surround the beam pipe in which electrons and positrons collided head-on. Magnets guiding one of the beams towards the collision point are visible on the left of the photograph. (University of California, Lawrence Berkeley Laboratory.)

produces a muon and an antimuon; so, wondered Perl and his colleagues working on the Mark I, could the collisions produce a third, heavier charged lepton, along with its antiparticle? And if so, how were the physicists to recognise it?

They knew that a new lepton would have to be relatively heavy, otherwise it would already have been discovered; indeed, the implications were that it would have to weigh more than 10 times the muon. This meant that any new lepton would decay in a variety of ways to different combinations of the relatively lightweight pions, which, in turn, meant that the new particle's lifetime would be very short, less than a billionth of a second. So, the physicists would be able to find it only through the patterns of particles it produced in its decays.

The decays to pions would be difficult to spot, because there would be many other ways of creating pions. However, assuming the lepton family repeated itself in a third heavier 'generation', it was possible that the new lepton would decay simply to a muon plus a muon-antineutrino and a new neutrino, or that it would produce an electron together with an electron-antineutrino and a new neutrino. The antilepton produced simultaneously in the annihilation could also decay in similar ways. So Perl and his colleagues decided to search for occasions in which an electron and an antimuon (or *vice versa*) appeared simultaneously, with nothing else, for the neutrinos would leave the scene undetected. As with the original beta-decay studies, these events would appear not to conserve energy, as a large amount would disappear with the neutrinos.

In 1974, they were rewarded with their first handful of such events, but it took the team two years of careful study to find more events and convince themselves that were indeed observing the decay of a third type of charged lepton. Perl recalls:

When we began the search we didn't know whether heavier leptons existed; and we didn't know whether we were looking in the right mass range. Thus it was to our surprise, and I think to the greater surprise of other particle physicists, that we found a new electrically charged lepton . . .[16]

Perl's team gave the new particle the name *tau*, from the initial letter of the Greek word 'triton', meaning 'third'.

Since these early findings, the existence of the tau has become well established. It is nearly twice as heavy as the proton, or some 3500 times heavier than the electron, and 18 times as heavy as the muon. It is very short-lived, with a lifetime of three tenths of a million-millionths of a second. But what of the hypothesised tau-neutrino? Certainly the tau can decay in such a way that it emits only one neutrino, along with a few charged pions; is this neutrino really different from the electron-neutrino and the muon-neutrino? Does it really carry some property of 'tau-ness'?

The answer is that we do not know for sure. Physicists have yet to do a 'three-neutrino' experiment – an experiment that demonstrates that neutrinos produced in the decays of taus can produce only taus, and not muons or electrons. However, we can tell from the rate at which certain particles decay that the neutrino emitted by the tau cannot be exactly the same as the electron-neutrino or the muon-neutrino. And the electroweak theory, which works so well in describing the electromagnetic and weak interactions of particles, assumes that there is a tau-neutrino. So the very success of the electroweak theory provides strong circumstantial evidence that the tau-neutrino does indeed exist. Moreover, as we shall see later in this chapter, 1989 brought some new results which showed that nature certainly seems to require three types of lightweight neutrino – no more, no less.

Weighing neutrinos – II

It is natural to guess that the neutrino masses will roughly parallel the corresponding charged lepton masses.[17]

Steven Weinberg, 1986.

The three charged leptons show a remarkable range in masses, from the electron at 0.511 MeV to the muon at 105 MeV to the tau at 1784 MeV. We know that the mass of the electron-neutrino is by contrast very small, being

Tracks of a muon and an electron
in the Mark I detector, with
nothing else, reveal the decays
of a third type of lepton, the tau.
A tau and an antitau produced in
an electron–positron annihilation
at the centre of the detector have
decayed, one to a muon and two
neutrinos, the other to an electron and two neutrinos. The neutrinos leave the scene undetected. (SLAC.)

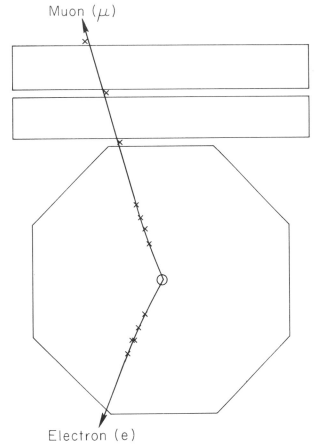

almost certainly no more than a few electronvolts if it is not actually zero.
(This is the neutrino measured in the tritium beta-decay experiments
described in Chapter 3). But what of the muon-neutrino and the tau-
neutrino? Is it possible that they have larger masses than the electron-
neutrino, while still being of far smaller mass than their charged associates,
the muon and the tau?

A team at the Swiss Institute for Nuclear Research (now the Paul Scherrer
Institute) in Zürich has made the most precise measurement of the mass of
the muon-neutrino. The principle behind the experiment is simple, but the
accuracy it requires makes it very difficult to perform.

The technique is to observe the decay of a charged pion at rest in the
laboratory. The pion decays into two particles – a muon and a muon-
neutrino – which should move away with equal and opposite momentum,
their total mass-energy being equivalent to the mass of the pion. Since we
know the mass of the muon and the mass of the pion, we can calculate what
the momentum of the muon should be if the neutrino's mass is zero. Any

deviation of the measured momentum from this calculated value should then reflect the mass of the neutrino.

The difficulty with the experiment lies in the relatively large amount of energy released in the pion's decay. The charged pion is around 34 MeV heavier than the muon, and if the neutrino is assumed to have zero mass, then the muon's momentum, expressed in the same units, will be almost 30 MeV. The team at SIN succeeded in measuring the muon's momentum to a very high degree of accuracy – 0.0028% – and showed that at this level the measured momentum agrees with the momentum calculated for a neutrino of zero mass.

But, perhaps surprisingly, this does not allow the team to say that the mass *is* zero. For if you assume a neutrino mass of 0.25 MeV, then you find that it changes the calculated momentum for the muon by only 0.00082 MeV or 0.0028% – in other words, by an amount comparable to the measurement errors! So, the best that the team can say is that they believe the muon-neutrino's mass to be less than 0.25 MeV, with a confidence level of 90%. This value may be much larger than the limit for the electron-neutrino's mass, but it is much smaller than the mass of the muon itself. The result therefore meshes well with the picture of neutrinos being very much lighter than the charged leptons.

With the tau-neutrino, the difficulties in measuring the mass precisely become even greater, for the difference between the mass of the tau and particles such as the muon or the pion is very large – in the region of 1600 MeV. The best results have come from observing the decay of the tau into a tau-neutrino and five charged pions (two positive pions and three negative pions in the case of the negative tau), where the five pions can be treated together as if they were a single particle with a mass that can approach that of the tau.

This technique has been exploited by the team working with the detector called ARGUS at the high-energy physics laboratory, DESY, in Hamburg. The ARGUS detector is located at an electron-positron storage ring, known as DORIS II. The ring is similar to SPEAR, but operates at total collision energies of around 10 GeV. The detector itself resembles the Mark I, with which the tau was discovered, but employs more modern, sophisticated apparatus for detecting and measuring the particles produced in the annihilations.

The electron–positron collisions at DORIS can produce tau–antitau pairs. To measure the mass of the tau-neutrino, the team on ARGUS search for event in which the tau decays to five charged pions and a neutrino, while the antitau decays in the simplest possible fashion to a single, visible charged particle plus one or more invisible neutrinos. Such decays are rare, occurring only about five times for every 10000 tau decays. Out of some 2 million events collected between 1983 and 1986, in each of which the electron–positron annihilations produced several pions, the ARGUS team found only 11 examples that fully satisfied all their criteria for five-pion decays of the

The ARGUS detector at the DESY laboratory in Hamburg. Physicists working with this detector have produced the best measurements of the mass of the tau-neutrino. (DESY.)

tau. But this small sample was sufficient to make a big impact on the upper limit for the tau-neutrino's mass.

For each of these 11 events, the researchers measured the momentum of the five charged pions in the tau's decay, and deduced not only the total momentum of the pions but also their total energy. In we define our units appropriately, then:

$$(\text{mass})^2 = (\text{energy})^2 - (\text{momentum})^2$$

so the team could also calculate a 'five-pion mass' for each event, by treating the five pions as if they were in effect one particle.

It so happens that the number of events that occur with a 'five-pion mass' close to the tau's mass (the maximum possible) depends in a very sensitive manner on the mass of the tau-neutrino. And in two of the eleven events detected in ARGUS, the five-pion mass was very close to the mass of the tau. By comparing the numbers observed to the numbers calculated the researchers could conclude, with a 95 per cent confidence level, that the mass of the tau-neutrino must be less than 35 MeV. Again this figure seems large compared with the limit for the electron-neutrino's mass, but it is small compared with the mass of the related charged lepton, the tau.

It should be possible for researchers to reduce this limit for the tau-

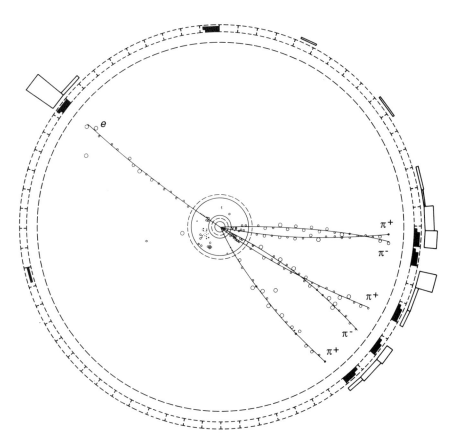

Tracks in the ARGUS detector shows the decay of a tau and an antitau produced in an electron–positron annihilation at the centre. In this case, the antitau has decayed to a tau-antineutrino and five pions, while the tau has decayed to an electron and an electron-antineutrino and a tau-neutrino. By measuring the effective mass of the five pions, the ARGUS team has found that the mass of the tau-neutrino must be less than 35 MeV. (DESY.)

neutrino's mass, not only through collecting more events of this kind, but also through improving the measurement of the mass of the tau itself. This is a major source of uncertainty in calculating the five-pion mass spectrum for comparison with the measurements. Experiments to take the limit on the tau-neutrino's mass to below 10 MeV would be possible at a 'tau-charm factory' – a high-intensity electron–positron collider designed to produce large numbers of tau particles, along with particles that have a similar mass to the tau, but which contain the heavy variety of quark known as 'charm'. Such a facility has been proposed by several groups of physicists, to enable a whole range of high-precision experiments, but so far no country or institution has committed itself to building a machine of this kind.

Oscillating neutrinos

> We do not know . . . [if] neutrinos are massive or massless. We do not know if the potentially massive neutrinos are Majorana or Dirac, and we do not know if these neutrinos can oscillate among flavours. . . . In short, there is a great deal we do *not* know about neutrinos.[18]
>
> *Jeremy Bernstein, 1984.*

Maurice Goldhaber once remarked that neutrinos 'induce courage in theoreticians and perseverance in experimenters';[19] they should certainly not induce complacency in anyone. In following the history of neutrinos, there is often the feeling that just when you think you are safe in your knowledge, somebody comes along and pulls the metaphorical rug from under your very feet, so that you begin to doubt what you think you know, and indeed what you know you think. Lewis Carroll would have loved neutrinos!

There is, after all, so little we really know about neutrinos, and what little we do know raises questions that are difficult to test. Some of the puzzles about neutrinos stem from one of the few known facts – their zero electric charge. This leads to the question of whether the neutrino is identical to its antineutrino (see Chapter 3). It also raises the possibility that a neutrino of one type can change into another type – a phenomenon known as 'neutrino oscillation'. In other words, are the objects that we have been calling the electron-neutrino, muon-neutrino and tau-neutrino really unrelated fundamental particle states?

The idea that a particle can oscillate between different states goes back to the mid-1950s, and studies of *kaons* – particles that were first found in cosmic-ray interactions in the late 1940s. In 1953, in a scheme for classifying the various newly-found particles, theorist Murray Gell-Mann represented the neutral kaon, K^0, and its antiparticle, $\bar{K}^0$, as two distinct particles. This proposal led Enrico Fermi to ask how you might tell the neutral particle and antiparticle apart. Both the K^0 and the $\bar{K}^0$ can decay to a pair of pions, one positive and one negative to conserve electric charge. So, if you observe this decay how do you know which particle has originated it – the K^0 or the $\bar{K}^0$? The integrity of the neutral kaon as a well-defined particle is clearly at stake.

In searching for an answer, Gell-Mann and fellow theorist Abraham Pais came up with a startling new idea, 'so unfamiliar that it was thought best not to report it at the Glasgow conference [in July 1954].[20] The key to the problem lay in realising that what we observe must be a *mixture* of the two states, K^0 and $\bar{K}^0$. To describe this phenomenon, Gell-Mann and Pais turned to a well-established principle of quantum mechanics – that of *superposition*. They combined the quantum mechanical expressions for the K^0 and the $\bar{K}^0$ in the two ways prescribed by superposition, and found that the results corresponded to quantum states with definite but distinct lifetimes. One state could decay to two pions, as had already been observed, while the other could decay only to three pions, but with a much longer lifetime, as it is more difficult to make three particles rather than two.

Pais and Gell-Mann argued that these states of definite lifetimes must correspond to the observable particles, while the K^0 and the $\bar{K}^0$ are the basic particle states of definite mass. Two years later, in 1956, the existence of the long-lived state, known as K_{Long} in contrast to K_{Short}, was confirmed, when Leon Lederman and colleagues at Brookhaven observed the decay of a neutral kaon to three pions for the first time.

Inspired by the work of Gell-Mann and Pais, Bruno Pontecorvo in Dubna turned to consider the possibilities of quantum mechanical mixing in another neutral particle – the neutrino. In 1957 he first discussed the idea of neutrino–antineutrino mixing; ten years later he turned to a different idea, that through mixing the electron-neutrino could change into a muon-neutrino and *vice versa*. Ziro Maki and colleagues at Nagoya University, and Yasuhisa Katayama and others at Kyoto University, also made similar proposals in 1962.

The basic idea behind neutrino mixing is that the states that participate in weak interactions are not exactly the same as the states – if they exist – of specific mass; instead, the weak interaction states, the particles we call the electron-neutrino, the muon-neutrino and the tau-neutrino, are mixtures of states of specific mass. If this is the case, then when electron-neutrinos, for example, are emitted at a certain energy, the different mass states will propagate through space at different velocities. As the particles move through space, the mass states will then become out of phase with each other, so that the mixture they form will change with time.

In this way, the mixture that originally corresponds to the particle we call the electron-neutrino, evolves into the mixture corresponding to the muon-neutrino or the tau-neutrino. And a beam that starts out as a pure beam of electron-neutrinos, will after some time contain muon-neutrinos and tau-neutrinos. The actual proportions of the different neutrino types will depend on only a few parameters – the energy, the distance from the source of the beam and, importantly, the difference in the masses-squared (m^2) of the base states.

Because neutrinos are neutral, such neutrino oscillations would conserve electric charge. Moreover, the number of leptons would remain the same. All that would happen is that the type or *flavour* of neutrino would change.

But does nature allow such changes? The answer is that we do not know for sure.

The electroweak theory, which unites electromagnetic and weak interactions, does not demand that leptons always maintain their flavour. It is observation that dictates this 'law', as in the 'two-neutrino' experiment, where muon-neutrinos produced only muons, preserving 'muon-ness', the muon's flavour. However, in its simplest form the electroweak theory does not allow neutrinos to have mass, and so excludes the possibility of neutrino oscillations.

It is possible to modify the electroweak theory slightly, by introducing very small masses for the neutrino base states, but in an *ad hoc* manner that fails to explain the size of the masses. So, if neutrinos *do* oscillate, this may be an indication of effects that lie outside the electroweak theory. Attempts at 'grand unified theories', which aim to incorporate the strong force into the same mathematical framework as the electromagnetic and weak forces, often give rise to flavour-changing reactions between leptons, albeit at a very low level.

The interest in grand unified theories was one of the reasons for a surge of interest in neutrino oscillations in the 1980s. Another reason is that if electron-neutrinos change flavour between the Sun and Earth, then we might be able to explain the apparent short-fall in the number reaching Earth (see Chapter 6). A third reason is that it turns out that observable oscillations could occur for very small masses – much smaller than we can hope to measure directly.

There are two basic ways to search for neutrino oscillations: through the disappearance of a neutrino flavour or through the appearance of one. A typical 'disappearance' experiment is to detect electron-neutrinos produced in a nuclear reactor, through the process of inverse beta-decay. If you detect fewer reactions than you expect, then a possible explanation is that some of the electron-neutrinos have changed to muon-neutrinos or tau-neutrinos between the reactor and the detector. Another possibility is to detect muon-neutrinos produced at an accelerator, and look for any reduction in the number of interactions observed. Ideally, in either case, you should perform the experiment at two or more distances from the source of the neutrinos, as any real oscillations should vary periodically with distance. Moreover, many uncertainties, for instance in the number of neutrinos with a given energy, tend to cancel in comparing the results at the different positions.

'Appearance' experiments are possible only with neutrinos of energies high enough to produce the appropriate charged lepton. It would not be possible, for example, to observe the production of a muon using neutrinos from a reactor, even if electron-neutrinos could change to muon-neutrinos, as the neutrinos would not have enough energy to produce a muon. So appearance experiments fall generally in the domain of research at particle accelerators, although there has been some work on muon-neutrinos produced in the high-energy reactions of cosmic-rays.

What do such experiments show? Although there has been the occasional, headline-making indication of neutrino flavour-changing, the general consensus is that neutrino oscillations have not been observed in either appearance or disappearance experiments. But that is not to say that there is no evidence for neutrino oscillations. On the contrary, in 1990, not only did experiments studying neutrinos from the Sun hint at oscillations, as we shall see in Chapter 6, but so too did laboratory experiments measuring none other than nuclear beta-decay.

Ten years previously, in April 1980, Fred Reines and colleagues working at the Savannah River reactor, had announced that they had found signs of neutrino oscillations in their data. Further investigations at various reactors eventually proved there to be no discernible effect in reactor neutrinos, but in the ensuing months, two papers appeared in *Physics Letters*, indicating how it might be possible to discover evidence for oscillations in measurements of nuclear beta-decay. As Fermi had shown, back in 1934, the shape of the high-energy end of the spectrum for the electrons emitted in beta-decay depends on the mass of the neutrino emitted. So if the electron-neutrino is a mixture of mass states, then the electron energy spectrum should be a mixture of different spectra, each corresponding to the relevant mass states. The net effect is to produce deviations in the electron spectrum, which should be detectable if the amount of mixing is at the level of a per cent or more.

Spurred on by these ideas, J.J. Simpson at the University of Guelph in Ontario, began to investigate the spectrum from tritium decay, an area in which he was already an acknowledged expert. At first he found no indications for neutrino mixing. Then, with improved detection electronics, he discovered signs that the tritium spectrum deviated from theoretical predictions below energies of around 1.5 keV. The endpoint of the spectrum occurs at 18.6 keV, so the data suggested the emission of a heavy neutrino with a mass 17.1 keV greater than the 'standard' electron-neutrino – in other words, a mass of about 17.1 keV. The amount of mixing required to explain the data was about 3 per cent.

Simpson's results, published in April 1985, caused a certain amount of upheaval, as theorists tried to find out if their theories could accommodate a 17-keV neutrino, and experimental groups tried to find further evidence. As time went by, a confusing mixture of results emerged from different experiments. One area of enquiry was to investigate beta-decay in a different type of nucleus; if mixing with a 17-keV neutrino occurs at all, it should be a general property of all beta-decays. The major problem, as with measuring the endpoint, is to find a nucleus that provides an adequate number of decays in the energy region of interest. A number of investigations of the beta-decay of sulphur-35 showed scant evidence for neutrino mixing. However, Simpson, together with his student Andrew Hime, found further signs not only in an improved experiment on tritium beta-decay but also in a study on sulphur-35.

The conflicting signals from the various experiments tended to make most physicists lose interest in the 17-keV neutrino. But then, late in 1990, two

The energy spectrum for the beta-decay of tritium, with the low-energy portion enlarged to show the effect discovered by J. J. Simpson at the University of Guelph. The solid line shows the theoretical prediction, which begins to deviate slightly from the data at an energy 17 keV below the endpoint. Simpson found that he could account for this deviation if he assumed that about 3 per cent of the time the tritium emits a neutrino with a mass of 17 keV. (*Physical Review Letters* **54**, 1891, 1985.)

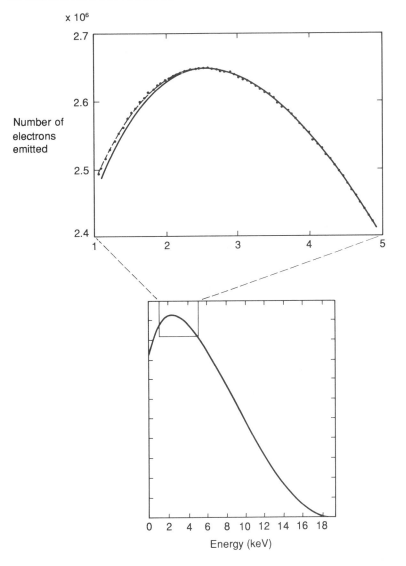

new experiments reported evidence for its emission. In one of the experiments, Andrew Hime, who had moved to Oxford University, and Nick Jelley, had studied sulphur-35 again, this time with apparatus designed to improve upon Simpson's spectrometer. Again they found evidence for a distortion in the beta-spectrum, corresponding to a mixing with a 17-keV neutrino at level of nearly 1 per cent. Meanwhile, Eric Norman and a team at the Lawrence Berkeley Laboratory in California had been investigating the beta-decays of carbon-13, and while they had accumulated nowhere near as much data as Hime and Jelley, they too found evidence for a 17-keV neutrino. And by the autumn of 1991, Hime and Jelley had found the effect in yet another nucleus, that of nickel-63.

All this was sufficient to rekindle interest in the possibilities of a 17-keV neutrino, even if it was not quite enough to convince fully a traditionally sceptical community of particle physicists. The new results led instead to a situation of 'wait and see'. While theorists could ponder over how they might fit a 17-keV neutrino into their theories, most physicists eagerly awaited further evidence. In particular, studies with magnetic spectrometers, of the kind described in Chapter 3 for measuring the mass of the neutrino in tritium beta-decay, have failed to produce any evidence for the 17-keV neutrino. Simpson, Hime and Jelley have used nonmagnetic spectrometers in their studies; they argue that there are systematic effects in magnetic spectrometers that are not fully understood, and which may obscure signs of the heavier neutrino. However, if the 17-keV neutrino is eventually clearly seen in completely different experiments on several different nuclei then it will be here to stay – and it will be one of the major discoveries that takes particle physics beyond the era of electroweak unification.

Counting neutrinos

Counting the number of massless neutrinos in terrestrial experiments has been an extremely difficult task. The upper limit [of] 6000 . . . was for some time the most successful result.[21]

Francis Halzen and K. Mursula, 1983.

One, two, three – just how many different types, or flavours, of neutrino are there? Once again, theoretical physics does not provide much of a guide in helping with what we do not know about neutrinos. The best tools theorists have at their disposal at present are provided by what has become known as the 'Standard Model'. This consists of two parts: the electroweak theory of weak and electromagnetic interactions, and the theory called quantum chromodynamics, which deals separately with the strong interactions of quarks.

According to the Standard Model, the basic particles – the leptons and the quarks – are related in pairs. Thus the electron is partnered by the electron-neutrino, the muon by the muon-neutrino and the tau by the tau-neutrino. A similar pairing occurs among four of the known quarks – *up* with *down*, *charm* with *strange* – and, indeed the model works so well that most physicists firmly believe that a sixth quark, *top*, exists to pair with *bottom*, and that is only a matter of time before it is discovered. In this way, the pairs of leptons are matched by pairs of quarks, of steadily increasing mass, each pair belonging to one 'generation' in each family of particles.

Does the pattern stop with three generations per family, or are there more? The Standard Model begs the question: it cannot say, although there are some general arguments as to why the number of generations cannot be too great. And, until recently, experiments in particle physics were even more useless than theory at answering the question. The best indications came from cosmological arguments about the formation of matter in the early Universe. As Chapter 7 describes in more detail, the relative proportion of the two lightest elements, helium and hydrogen, observed now in the Universe suggests that no more than four generations of lightweight (or massless) neutrinos have ever existed. However, in the autumn of 1989 measurements in particle physics finally showed once and for all that there are only three flavours of neutrino – the tau's the limit!

The answer to this question that had tantalised particle physics at least since the discovery of the tau in the mid-1970s lay not with leptons nor with quarks, but with a third type of particle – a force-carrying particle called the Z^0, which is generally pronounced 'zed-nought' in Europe, 'zee-zero' in the US! The Z^0 is like a heavy photon; indeed, a very heavy photon as its mass is nearly hundred times the mass of the proton. Just as the photon carries the electromagnetic force between charged particles in a game of 'quantum catch', so the Z^0 carries the weak force between particles, as we shall see in Chapter 5.

The Z^0 is arguably the major star of electroweak theory, the branch of the Standard Model that links the electromagnetic and weak forces. The particle was predicted by the theory more than a decade before its eventual direct detection in 1983. By studying its properties in detail, experimenters hope to find subtle clues that lead to a theory beyond the Standard Model – a theory that is more comprehensive and which leaves fewer questions unanswered. However, because it is so much heavier than other particles, prior to 1989, experimenters had only ever produced about 50 examples of a Z^0. Then in the late summer of 1989, results based on many more Z^0 particles from three different laboratories began to emerge.

Thanks to a daring and ingenious modification to its 20-year-old electron accelerator, the Stanford Linear Accelerator Center, in California, was able to produce about 500 Z^0s during the summer of 1989. The modifications allow the basic machine to accelerate not only electrons but also positrons to energies as high as 50 GeV. At the end of the linear accelerator, the electrons and positrons diverge along two arcs of magnets, which bring the particles round to meet head-on. The resulting collisions have a total energy of 100 GeV, enough to produce the massive Z^0 particles. Also in the US, physicists working at Fermilab in Illinois announced in July 1989 that they had analysed the decays of some 500 Z^0s. These were produced in the high-energy collisions of protons and antiprotons in the Tevatron, Fermilab's circular machine which accelerates the two beams to 900 GeV before they collide. However, the results from both American laboratories were soon to be eclipsed by those from a mighty new accelerator, designed to be a veritable production line of Z^0s – a 'Z^0 factory'.

CERN, the European laboratory for research in particle physics, straddles the border between France and Switzerland a little to the north to Geneva. It began life in the 1950s, its work centred on a proton accelerator. But early in the 1980s the decision came to build a huge new machine, called LEP for Large Electron–Positron collider. LEP was designed to take to the practical limits the concept of electron–positron colliders, which had first proved so fruitful with SPEAR, where the tau and several new particles containing charmed quarks were discovered during the 1980s. It was also designed to be a Z^0 factory.

When electrons and positrons annihilate, their energy can rematerialise as any kind of particle, or particles, consistent with basic conservation laws. So the total charge of the products must be zero, and the total mass-energy must be no greater than the combined energy of the annihilating pair. Experience dictates that the total number of leptons minus antileptons, and the total number of quarks minus antiquarks must also be zero, in keeping with the initial conditions. This last condition implies that particles and antiparticle are generally produced in pairs. However, the Z^0 is neither a quark nor a lepton, and moreover it is electrically neutral. Put together, this means that an electron–positron pair can annihilate to produce a single Z^0, provided there is enough energy, around 90 GeV, to create the particle's mass.

LEP, in its first phase, was designed to collide electrons with positrons at a maximum total energy of 100 GeV – just enough to make a production line for Z^0s. To achieve this the machine has been built on an unprecedented scale, with a ring of magnets 27 km in circumference, in an underground tunnel that sweeps out from CERN to the foothills of the nearby Jura mountains and back. This great size is necessary because high-energy electrons and positrons readily radiate energy as they follow curved paths. So to minimise the amount of energy lost in this way the ring must bend as gently as possible.

The successful completion of LEP was a tremendous achievement in engineering of all kinds – especially civil, electrical and electronic. The same is true for the huge detectors that monitor annihilations at four points where the electron and positron beams collide. Each detector is a coordinated masterpiece of thousands of parts which together trap as many as possible of the particles created in the annihilations. Only the wily neutrinos can escape direct detection. Yet as is often the case, through careful measurement of the other particles produced, the experimenters can learn a great deal about the invisible neutrinos.

One of the very first questions the experimenters at LEP hoped to answer concerned the number of neutrino flavours, for one key to resolving this lies in the decays of the short-lived Z^0. The Z^0 can decay to any lighter particles, provided that the usual laws of conservation of charge, mass-energy, total number of leptons and so on are not contravened. In general, the more ways there are for a particle to decay, the shorter its life, so the lifetime of the Z^0 should reflect the number of lightweight neutrinos. But how do you measure the lifetime of a particle like the Z^0, which decays almost as soon as it has been made?

The answer is that you measure its *width* – that is, the 'uncertainty' surrounding its mass. The quotation marks here are to indicate that I am not referring to experimental uncertainties due to limits in the precision of measurements. Rather, I am referring to an intrinsic variation in the mass of a short-lived particle.

Werner Heisenberg first demonstrated in 1927 that there are certain pairs

(a) DELPHI, one of the huge detectors at the Large Electron Positron Collider (LEP) at CERN, Geneva, that have made detailed studies of the decays of the neutral carrier of the weak force, the Z^0 particle. (b) The decay of a Z^0 at the heart of another detector, ALEPH, produces two well-defined sprays of particles in opposite directions. (CERN.)

of quantities that cannot be measured simultaneously with arbitrary accuracy, a phenomenon that becomes manifest only on the subatomic scale. One such pair comprises energy and time, and Heisenberg's 'uncertainty principle' implies that the shorter the lifetime of a particular energy state, then the less well-defined the energy. In terms of a particle such as the Z^0 this means that the shorter its lifetime, the broader the range of values that its mass can have – in other words, the greater its 'width'. Since LEP can make thousands upon thousands of Z^0s, the experiments can make a very accurate measurement of its range in mass.

To 'weigh' the Z^0, the experimental teams at LEP took measurements at various collision energies, stepping slowly from the low-energy side of the Z^0's mass to the high-energy side. A graph of the number of Z^0s produced at each energy shows that the number begins to rise at about 89 GeV, sweeping up to a peak at 91.1 GeV, and falling smoothly to a low level again around 94 GeV. The width of this peak at half its height gives a direct measure of the Z^0's average lifetime – and the key to discovering the number of neutrino flavours.

With a total of some 11 000 Z^0s gathered between the four experiments after only the first few weeks of LEP's operation, the physicists had an accurate enough measure of the width of the Z^0's mass peak to be able to comment on the number of ways the Z^0 can decay to neutrinos. The more flavours of neutrino there are, the shorter lived will be the Z^0 and the greater its width, each flavour adding 150 MeV to the final width. The result, of 2.58 ± 0.08 GeV, implies that the chances of there being an additional

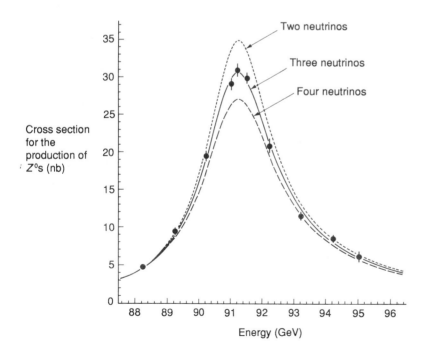

Measurements of the decays of thousands of Z^0 particles at LEP give a range in mass for this short-lived particle that is inversely related to its lifetime: the more ways the Z^0 has to decay, the shorter its life and the broader the width of this peak. The measured width indicates that there are three neutrinos to which the Z^0 can decay – no more, no less. (CERN.)

neutrino flavour beyond the three already known are less than one in a thousand.

Sixty years after Pauli first suggested the existence of a lightweight neutrino, we now know that there are three distinct types of such particle – no more, no less. True, there may be much heavier neutral particles, with similar properties, which experiments have still to reveal; such 'heavy neutrinos' often figure in the attempts of theorists to adventure beyond the Standard Model. However, in the realm of experiment, the Standard Model stands firm – its very success established in part by the invaluable role of the lightweight neutrinos as 'nuclear spaceships', as the following chapter reveals.

5

Nuclear spaceships

Neutrinos have yet to play the role which Rutherford designed for his alpha particles, but . . . there is still hope that the analogous explorations may take place.[1]

Leon Lederman, 1967.

THE TWO-NEUTRINO EXPERIMENT had proved that an experiment in a neutrino beam could not only work, but could indeed help to lift the curtain obscuring the basic nature of matter. However, it had resolved only one of the issues that had fired the original enthusiasm for neutrino experiments. In particular, there remained the question, 'What is the correct theory of the weak force?'

Enrico Fermi had given credibility to the neutrino when he built his theory for beta-decay in 1934. Later, his theory was shown to apply equally well to other 'weak' interactions and was modified only to incorporate the surprising discovery in 1957 that the weak force does not respect spatial, or parity, symmetry (see Chapter 3). Yet although Fermi's theory always gave the right answers, everyone knew that it could not be completely correct.

The problem lay with the theory's predictions for the rates of reactions – the cross-sections. According to Fermi's theory, the probability that an energetic neutrino interacts with an electron, for example, rises in proportion to the neutrino's energy. This leads eventually to the nonsensical situation where the neutrino has a greater than 1 in 1 chance of interacting with the electron! In other words, although Fermi's theory works perfectly adequately at the low energies characteristic of processes such as beta-decay, it is doomed eventually to fail at higher energies. But what kind of theory *will* work at high energies?

One key to resolving this 'energy crisis' is as old as Fermi's theory itself, and had, in fact, been contained in the attempt of Fermi's contemporary, Hideki Yukawa, to describe the weak force. Yukawa had based his work on the quantum description of the electromagnetic force, in which charged

In Fermi's theory of beta-decay the neutron simultaneously turns into a proton and emits an electron and an antineutrino at one point in space and time. But this theory is destined to give the wrong answers at high energies. Instead, it must be replaced by a theory of the weak force in which the neutron transmutes into a proton by emitting a weak carrier particle, the W^-, which, in turn, decays into an electron and an antineutrino – rather as Yukawa proposed (see p.73.)

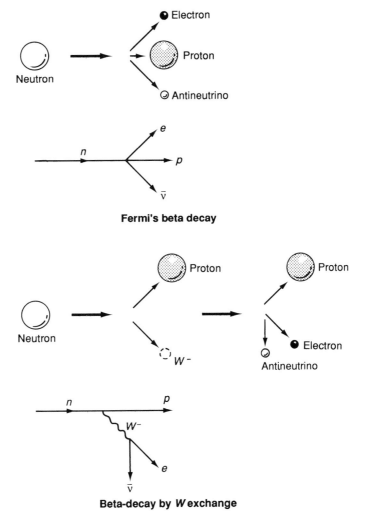

Fermi's beta decay

Beta-decay by W exchange

particles interact through a game of 'catch' a they pass photons between themselves. He managed successfully to describe the strong force between protons and neutrons in the nucleus in terms of a new game of catch in which the ball is a particle roughly 200 times as massive as an electron – the particle now known as the pion. But the same particle would not serve to describe the weak force, a possibility that Yukawa considered.

However, the quantum theory for the electromagnetic field proved to work so well that it remained a blue-print for a description of the weak force. Theoretical physicists continued to work on the concept of a game of 'weak catch' carried by an appropriate 'ball', in other words a 'weak photon', which became denoted by the letter W. A consequence of such an approach was that it promised to alleviate the ills of Fermi's theory, although it would not provide a complete cure.

Theorists knew that the particle to carry the weak force, the W, must be relatively heavy. This much was clear from the very success of Fermi's theory, which works so well at low energies. Fermi had assumed that the beta-decay of a neutrons occurs at a single point, the products of the decay emerging from exactly the point at which the neutron disappears. So at low energies, where the theory works, weak interactions must occur as if at a point; it is as if you do not 'see' the exchange of the carrier particle.

The implication of this low-energy behaviour is that the weak force is short-ranged, in contrast to the electromagnetic force, which has an infinite range. Photons, the carrier particles of the electromagnetic force, have zero mass, and quantum theory allows them to travel large distances. The short range of the weak force, on the other hand, implies that the W particles that transmit the force must be heavy, as the range of a force is smaller, the greater the mass of the carrier. Quite how heavy the W particle should be was not, however, clear.

The existence of a heavy weak carrier was not going to solve the crisis in Fermi's theory, but it would delay its onset. At higher energies, the exchange of an intermediary particle would modify the weak interaction and slow down the rate at which the reaction probability increases with energy. The weak interactions would no longer appear to occur at a single point, as in Fermi's theory, but would behave as if acting over a very small distance. In a sense, experiments at higher energies would be providing a microscope to reveal details that are unobservable at lower energies.

It was also possible that high-energy neutrino experiments might create real W particles, rather as the high-energy collisions of protons and neutrons liberate pions, the carriers that Yukawa predicted for the nuclear binding force. As it happens, neutrino interactions did not reveal the W particle. That honour fell to a different breed of experiment, which liberated W particles for the first time in 1983, and showed that they are some 90 times as heavy as the proton. But neutrinos did lead to a fully workable theory of the weak force, which could predict the mass of the W and therefore guide experimenters in the right direction.

Neutrino beams

> . . . during a decade experimental neutrino physics will pass from being technically unfeasible to one of the most important experimental fields in high energy physics.[2]
>
> *Colin Ramm, 1966.*

In the aftermath of the two-neutrino experiment, the challenge to experimenters was to 'scale up'. The number of neutrino interactions in the spark chambers at Brookhaven had been very small. To have a chance of detecting the W particle, experimenters would have to construct neutrino beams of higher intensity and build even larger detectors.

A technique to yield more neutrinos was, it happens, immediately to hand at CERN, the research laboratory near Geneva that had been set up in the

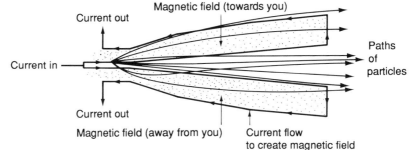

The first magnetic horn, built at CERN and photographed during assembly, was designed by Simon van der Meer to produce a more intense neutrino beam by focussing the pions and kaons before they decayed to produce neutrinos. The diagram shows how electric currents in the horn set up a magnetic field to focus the paths of the electrically charged parent particles. (CERN.)

1950s to serve particle physicists in Europe. Early in 1961, Simon van der Meer, a Dutch engineer, had published an internal report entitled 'A directive device for charged particles and its use in an enhanced neutrino beam'. In this report, van der Meer had described his ingenious idea for the device that has become known as the 'neutrino horn'.

One problem with forming a neutrino beam at an accelerator lies in the way the parent particles are produced when the high-energy beam from the accelerator strikes a target. These parents – charged pions and kaons – emerge at a variety of angles to the original beam direction, and give rise to a neutrino 'source' that is like a lamp that sprays light over a wide area. This effect is exacerbated by the several tens of metres that the charged particles must travel in order to produce enough decays to provide a reasonable number of neutrinos. It is not possible to focus the neutrinos themselves, because they are electrically neutral particles and so are unaffected by magnetic fields. So van der Meer realised that a device was needed to focus the parent pions and kaons. This would concentrate the spray of resulting neutrinos into a more parallel beam, rather as the lenses on a lighthouse concentrate the light from the central lamp.

Van der Meer's solution worked like the polished inner surface of a cone, which can direct light into a parallel beam through successive reflections. To bend the paths of the electrically charged pions and kaons, he used a magnetic field set up between two metal cones, one within the other. An electric current passing through the cones set up the magnetic field, which focussed the charged particles produced in a long thin metal target located in the neck of the horn. An added bonus was that the device would focus either positive parents or negative parents, depending on the direction of the current, and hence the magnetic field. This meant that it would produce an almost pure beam of neutrinos (from positive parents) or antineutrinos (from negative parents).

The need for van der Meer's 'magnetic horn' became clear in the middle of 1961, when the original ambitious plan for neutrino experiments at CERN was brought to an abrupt halt. Guy von Dardel discovered that the envisaged arrangement for using the beam in the proton accelerator (the Proton Synchrotron, or PS) produced far fewer pions than had been expected; the resulting low number of neutrinos would make the experiments impossible. The physicists at CERN had no alternative but to change tack, as it was already clear that the team at Brookhaven was much closer to making the first studies with neutrino beams. The new plan was to establish a high-intensity neutrino beam at CERN that would enable more detailed experiments than would be possible at Brookhaven.

The magnetic horn was a vital ingredient of CERN's new plans, but it required a second technical innovation. The parent particles had to be produced in the 'throat' of the horn, and this implied that the proton beam had to leave the accelerator to strike a target located at the narrow end of the horn. Fortuitously, in December 1959, Berend Kuiper and Gunther Plass had already devised a scheme for 'fast ejection' in which they could direct all the particles out of the accelerator in less than 2.1 microseconds – the time for one revolution around the accelerator ring.

CERN conceded the first prize in high-energy neutrino experiments to Brookhaven in 1962. By mid-1963, with the combined use of fast ejection and the neutrino horn, the European laboratory could boast the world's most intense neutrino beam. The next step would be to build an equally impressive detector.

The giant's mother

What kinds of instrument are required to fulfill our dreams of making detailed studies of neutrino interactions?[3]

Barry Barish, 1973.

In 1953, Donald Glaser, a young physicist at Michigan University, demonstrated the feasibility of a new kind of particle detector – the bubble chamber. His detector was not much more than a glass phial containing 3 centilitres of diethyl ether, an organic liquid. Over the following years, however, bubble chambers grew, particularly under the inspired direction of Luis Alvarez at the Lawrence Berkeley Laboratory in California. He pioneered their large-scale use in the 1960s with the '72-inch' (1.8 metre) chamber, filled with liquid hydrogen, which made several discoveries. Then in the early 1970s, a giant bubble chamber called Gargamelle, containing 12 000 litres of liquid, began to make the dreams of detailed studies of neutrino interactions come true.

The original Gargamelle was the creation of France's great writer of the sixteenth century, François Rabelais. She was the mother of Gargantua, the giant whose exploits gave Rabelais the opportunity to satirise the state of contemporary France. The modern Gargamelle, the bubble chamber, was the creation of a team of French physicists and engineers led by André Lagarrigue. This giant gave physicists a window not only onto new behaviour of the weak force, but also onto the structure that lies within the protons and neutrons of the atomic nucleus.

In a bubble chamber, the tracks of electrically charged particles that have passed through a liquid become visible as trails of tiny bubbles. The technique is to keep the liquid under pressure, releasing the pressure only

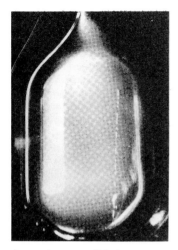

A cosmic-ray particle leaves a fine trail of bubbles, illuminated by a flashlamp, in the first bubble chamber – a glass phial only 3 centimetres long, containing diethyl ether. (D. Glaser, University of California.)

fractions of a second before a charged particle enters the chamber. The sudden release in pressure makes the liquid 'superheated': although still a liquid, it is now at a temperature above its boiling point for the reduced pressure. But if a charged particle passes through the liquid, it loses small amounts of energy as it ionises atoms along its trail, and this energy is just enough to initialise boiling in the unstable, superheated liquid. The result is a chain of bubbles along the particle's path.

Bubble chambers became popular detectors during the 1950s, because they can provide a great amount of information about reactions between particles that enter the chamber in an energetic beam, and the atomic nuclei in the liquid. Any sluggish charged particles that are produced leave dense trails of bubbles, while swift particles form fine tracks. When the bubble chamber is set between the poles of a large electromagnet, the tracks of the particles bend in the magnetic field. They curve in opposite directions according to the positive or negative charge of the particles, and the amount of bending reveals the momentum of the particles. Moreover, if the chamber contains liquid hydrogen, the particles in the beam react simply with single protons, the nuclei of hydrogen.

Neutral particles are a problem, however, because they do not leave ionised trails. One possibility is to fill the chamber with a liquid in which neutral particles are more likely to interact and produce some characteristic pattern of tracks – in other words, a 'signature' for the neutral particle. This is the case with gamma-rays, which can convert into pairs of electrons and positrons in a process that is the reverse of electron–positron annihilation. The electron and positron are oppositely charged, so the gamma-ray's signature in a bubble chamber is a pair of tracks curving in opposite directions in the magnetic field. When the electron and positron have low momentum, their tracks curl tightly, like a set of ram's horns.

Gamma-rays are more likely to 'convert' in the electric field around the highly charged nucleus of a heavy element, so to make gamma-rays more visible in a bubble chamber it can pay to fill the chamber with a 'heavy' liquid – a liquid with a large positive charge on its nuclei. One liquid with appropriate properties is Freon, a compound of carbon, fluorine and bromine (CF_3Br), which is better known nowadays as a bane of the Earth's ozone layer. Freon is more than 200 times denser than liquid hydrogen, which means that a gamma-ray will travel only one-hundredth the distance through Freon that it would through liquid hydrogen before converting into an electron–positron pair. However, the liquid extracts its price; the particles in the beam no longer react simply with single protons, as in liquid hydrogen, but with a tangle of protons and neutrons bound in complex nuclei.

When the possibility of creating neutrino beams arose in 1960, experimenters immediately considered using bubble chambers as detectors. The heavy-liquid bubble chambers that had only recently been developed were an obvious choice. At CERN, Colin Ramm led a team building a 500-litre chamber that could contain either propane or Freon, or a mixture. When the neutrino beam at CERN's Proton Synchrotron came into operation in mid-1963, Ramm's chamber was the first in line to receive the neutrinos. After a total of some 60 days of running, data from the chamber had confirmed the results of Brookhaven's two-neutrino experiment, with as many 454 events containing a negative muon and only 5 with a relatively energetic electron. The bubble chamber had proved itself as a neutrino detector.

But the numbers of events were still small. To make any progress in studying neutrino reactions would require 10–100 times as many events, and this implied a much larger detector. Among those who realised this was André Lagarrigue, a physicist at the École Polytechnique in Paris and one of the pioneers of heavy-liquid bubble chambers. His colleague, André Rousset, later recalled:

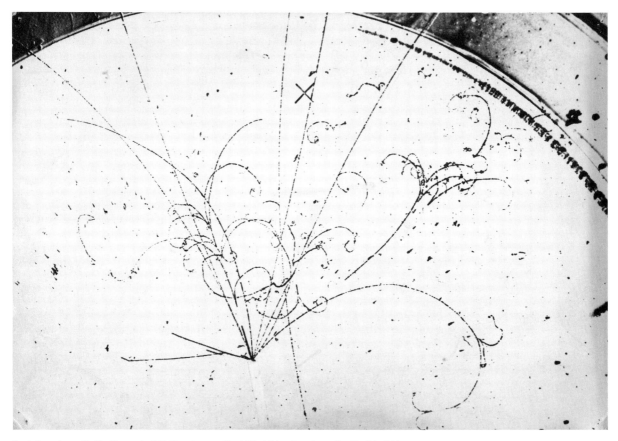

A picture from Colin Ramm's 500-litre heavy-liquid bubble chamber, the first bubble chamber to study neutrino interactions at CERN. Here a neutrino has interacted near the centre of the chamber, producing a spray of charged particles, which left trails of bubbles in the liquid. Notice how invisible gamma-rays were also produced, to be made visible ony when they converted in the heavy liquid into electron–positron pairs, which leave short tracks curling in opposite directions. (CERN/Courtesy G. Myatt.)

André Lagarrigue, from the École Polytechnique, who master-minded the building of Garga-melle, the huge heavy-liquid bubble chamber that made several important discoveries at CERN during the early 1970s. (CERN.)

The idea was conjured up for the first time in conversations in the cafés close to the Piazza del Campo at Sienna in 1963. At this meeting on particle physics, we were discussing the results on high-energy neutrino interactions . . . The heavy liquid bubble chamber appeared a good detector, but . . . it was necessary to produce a much bigger apparatus.[4]

And so Lagarrigue began to drawn up plans for a truly enormous new chamber – Gargamelle. His initial task was to convince his fellow physicists of the value of such a project; then the authorities who were to provide the money to build the chamber; those at the Centre d'Études Nucleaires at the Saclay Laboratory, where the chamber would be built; and finally those at CERN, where the chamber would be installed. Lagarrigue was indefatigable in his campaigning for Gargamelle, and ultimately successful. On 2 December, 1965, CERN signed an agreement with the Commissariat à l'Énergie Atomique (CEA), who were to oversee the construction of the chamber.

Gargamelle's final size was dictated largely by the money available and the cost of the electromagnet – the single most expensive item – within which the chamber would eventually sit. The bubble chamber's shape was dictated by the need to be able to identify particles produced in interactions. In particular, this meant that the chamber had to be long enough to distinguish the long tracks of muons from the similar but much shorter tracks of pions, which interact more readily.

In their final design, Lagarrigue and his team of physicists and engineers proposed a chamber that was to be like a flattened cylinder, 4.8 metres long and 1.85 metres wide, with a volume of 12 cubic metres, which would hold

(a) The body of the big bubble chamber Gargamelle arrives at CERN, July 1970. Two rows of four large 'port holes' for the fish-eye lenses are visible along the upper and lower lengths of the chamber. (b) The many apertures in the inner walls of Gargamelle are visible in this interior view. As well as the large holes for the lenses there are smaller holes for flashlights and for the expansion system that allows the pressure on the liquid to be alternately increased and decreased during operation. (c) Here Gargamelle is installed between the coils of a huge magnet to provide the magnetic field to bend the charged particles so that their momenta and electric charges can be determined. (CERN.)

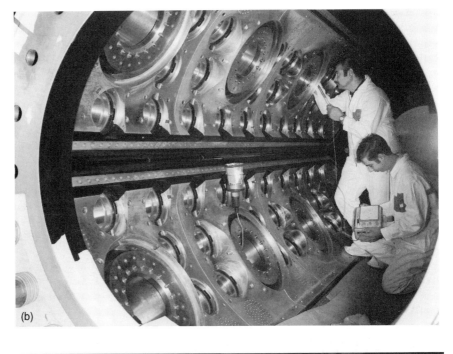

18 tonnes of Freon. They aimed to make as much of the chamber visible to cameras as possible, and decided to use two rows of four wide-angle 'fish-eye' lenses. This was intended to make 10 of the 12 cubic metres visible – an unusually large proportion for a bubble chamber.

Thus began what Paul Musset, one of the leading physicists in the project who shouldered much of the responsibility with Lagarrigue, has described as 'a beautiful technical adventure'.[5] After Lagarrigue's untimely death in 1975, Musset recalled the spirit of that adventure:

In the course of this work, the milestones of partial successes gladdened us and gave us the necessary spirit to follow the construction and tests. From time to time difficulties appeared and allowed us to gauge our technical limitations. Finally, this long enterprise was brought to its conclusion and I cannot forget the pleasure of André Lagarrigue when he could examine the first photographs and discuss the first technical improvements.[6]

Those first photographs appeared on the night of 8 December, 1970, during the first tests to expand the chamber, almost exactly four years after the agreement between CERN and the CEA. CERN's magazine, the *CERN Courier*, described the events as follows:

There was a tense atmosphere in the control room as the first thuds of the expansion cycle were heard, and everyone in turn peered through the observation window in an effort to see something. The optimists had a job to make out the tracks [due to cosmic rays]; the pessimists could see nothing at all. After half an hour the first film was completely exposed and developed and then at last the film gave an objective verdict. Cosmic ray tracks were there and the bubble chamber team joyfully celebrated their achievement.[7]

Less than a year later, the *CERN Courier* had something more to write about, when Gargamelle delivered her 500000th picture, in November 1971. Now Lagarrigue and his colleagues faced a different challenge: the careful analysis of all these photographs and more, which together contained important new evidence as to how the elusive neutrinos interact with matter.

Neutral currents

> Most of the breakthroughs in our field have come as a result of finding small numbers of events . . . I don't know if the one event I'm discussing is a great discovery. Let's hope it is![8]
>
> *Don Perkins, July 1973.*

Perkins, one of the leading figures in British particle physics, made these remarks to an audience of a new generation of aspiring high-energy physicists. Here, cloaked in Perkins' characteristically understated commentary, was history in the making. The one event came from Gargamelle: a lone event from the 375000 neutrino pictures and 360000 antineutrino pictures that had by this time been analysed.

The event did indeed prove to be a great discovery. It was the first example of a new type of neutrino interaction which confirmed a remarkable advance

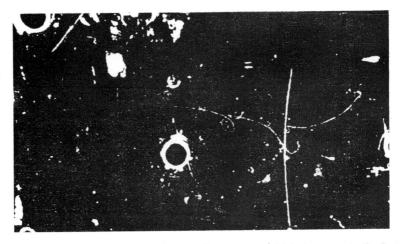

This portion of an image from Gargamelle captures an historic moment – the first example of a neutrino reacting with an electron via the weak neutral current. The track beginning towards the left of the picture is due to a lone electron, knocked from an atom in the liquid by an invisible high-energy neutrino. The electron's track displays the characteristic short curly branches due to interactions with other electrons in the liquid. (The white spots with black centres are some of the ring-shaped flashlights used to illuminate the trails of bubbles.) (CERN.)

in theoretical physics – the linking of the weak and electromagnetic forces within a single 'electroweak' theory. But if Perkins appeared equivocal in July 1973, he had every right to do so. Finding the event in question was like finding the needle in the proverbial haystack, and in this case the difficulty in proving that what you had found was indeed a needle and not a needle-like piece of hay made the task even more formidable.

Excitement over the possible 'unification' of the weak and electromagnetic forces had spread among the community of theoretical particle physicists late in 1971, sparked by a paper by a young Dutchman, Gerard 't Hooft. The basic idea was by no means new, having been developed during the 1960s by Abdus Salam in London and, independently, by Steven Weinberg at Harvard. But what 't Hooft had done was to bring electroweak unification alive by showing that he could rid the theory of nonsensical infinite results, a process known as 'renormalisation'. Sidney Coleman, a theorist at Stanford later wrote that 't Hooft had 'revealed Weinberg and Salam's frog to be an enchanted prince'.[9]

The reaction of 't Hooft's fellow theorists was enthusiastic as they swiftly checked his methods. However, as Weinberg himself pointed out, although the theory was beautiful,

What was not so clear was that our specific simple model was the one chosen by nature. That of course was a matter for experimentalists to decide.[10]

The stage was set for the experimenters to take over, and the theorists looked to them eagerly for the data that would make or break electroweak unification.

A key feature of the electroweak theory was its requirement of a new particle. This particle, called the Z^0, was an electrically neutral partner for the charged W particles that had long figured in attempted theories of the weak force.

If the Z^0 existed, then it would give rise to a distinctive game of 'weak catch'. No electric charge would change hands as the participants exchanged the Z^0. This contrasts with processes where a charged W particle is the 'ball', such as in the beta-decay of a neutron. Here, as a neutron turns into a proton, it emits a negatively charged W particle; the W^- then transforms almost immediately into an electron and an antineutrino. In a sense, the neutron changes its charge as it transmutes into a proton, handing the difference over to the electron. In analogy with electromagnetism, the different types of weak interaction are referred to as *charged currents* (where charge is exchanged) and *neutral currents* (where no charge exchange occurs).

Theorists quickly saw that the survival of the electroweak theory depended upon the existence of neutral currents, although the results of earlier experiments were discouraging as they had shown no sign of such processes. But the game was far from over, as the constructors of electroweak theory had a trump card to play. Salam and Weinberg could give a good estimate of what the relative strengths of the charged and neutral currents should be.

Late in 1971, Weinberg calculated that the earlier experiments had been barely sensitive to the predicted level of neutral currents, so 'there was every reason to look a little harder.'[11] And the best place to look for neutral currents was, in fact, in the interactions of neutrinos with matter. Neutrinos feel neither the electromagnetic force, as they are uncharged, nor the strong force, so the effects of weak interactions are not obscured by the much greater influence of these other forces.

Meanwhile, at CERN theorists began to alert their experimental colleagues to the significance of 't Hooft's work and to encourage the search for neutral currents. Gargamelle had already taken more than half a million pictures of neutrino and antineutrino interactions. Was the vital evidence buried in this mountain of film?

The neutrino experiment at Gargamelle was the responsibility of a team of more than 50 physicists from eight different European laboratories – in Aachen, Brussels, Milan, Orsay, Oxford, Paris (the École Polytechnique), London (University College) and at CERN. While a team of technicians looked after the day-to-day operation of the bubble chamber itself, the main role of the physicists was to organise the analysis of the photographs, and for this purpose the film was shared out between the various laboratories.

Gargamelle had two parallel rows of four wide-angle lenses to photograph as much of the interior of the chamber as possible. Two 70-millimetre films recorded the images, one film for each row of lenses. To study the images on the developed film, the physicists had to project them onto a special table, on which they could follow interesting tracks, while feeding measurements directly to a computer. A team from CERN and University College designed

In a typical weak *charged current* interaction a muon-neutrino emits a charged W^+ particle when it encounters an electron, and changes into a muon; the electron then changes into an electron-neutrino. In the related weak *neutral current* interaction, however, both initial particles retain their identities when the muon-neutrino emits a Z^0 particle, which is absorbed by the electron.

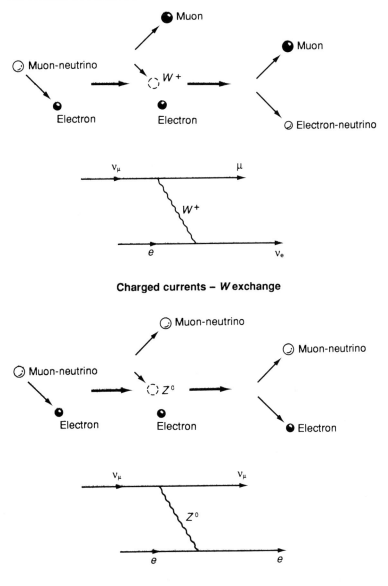

Charged currents – *W* exchange

Neutral currents – Z^0 exchange

and built a prototype projection system and measuring table, known as 'Gemini'. This formed the basis of 14 tables built by the Swedish company SAAB, which were distributed around the various laboratories.

As in many bubble chamber experiments, the first stage of the analysis of Gargamelle's pictures was not usually performed by the physicists, but by teams of 'scanners' – operators trained to pick out specific patterns of tracks and to make appropriate measurements. The scanners in this way filtered out 'events' with the desired characteristics, or 'signatures'. The physicists meanwhile formed small teams to investigate different aspects of neutrino scattering. They provided the scanners with rules for accepting events, checking them particularly when unusual or important signatures occurred. Much of the physicists' work, however, involved understanding how the wrong events could mimic the ones they were looking for, and being certain that they understood their 'filter' properly. They needed to be sure that they were looking at needles rather than hay.

Nowhere was this more true than in the search for neutral currents, which began in 1971 as simply part of the general program that the Gargamelle team was following. But after the great theoretical excitement aroused by 't Hooft's work had become apparent, the emphasis began to change. During 1972, the team began to put more and more effort into the search, with various subgroups concentrating on different kinds of neutral current.

One of the simplest examples of a neutral-current interaction is the 'elastic' scattering between a neutrino and an electron. In this case, the neutrino bounces off the electron and only energy and momentum are exchanged between the two particles. However, this process is very rare. A more common neutral-current process is the scattering of a neutrino from a nucleon (neutron or proton) bound in a nucleus. But this is more difficult to pin down experimentally.

Consider the problems involved. A neutrino enters a detector, leaving no track. It collides with something in the detector, and moves away at an angle, but still leaves no track. The only sign that anything has happened arises as the object the neutrino hits is set in motion, leaving some characteristic track, or pattern of tracks.

So, in the case of neutrino-electron scattering, a lone electron will apparently move off by itself. The most likely alternative way in which this can happen is for an *electron*-neutrino to collide with a neutron and instigate a form of inverse beta-decay – in other words, the collision produces a proton and an electron. If the proton is not given enough energy to leave a discernible track in the detector, the process will mimic elastic scattering from an electron. Although the Gargamelle team worked with beams nominally of *muon*-neutrinos (or antineutrinos), around 1 per cent (0.1 per cent) of the beam consisted of *electron*-neutrinos (antineutrinos) produced in the decays of neutral kaons. So the unwanted inverse beta-decay could occur, and make life difficult for the experimenters.

In neutrino–nucleon scattering the problems are more severe. Here the nucleon may be given enough energy to radiate pions, which leave a spray of tracks in the detector. With a beam of muon-neutrinos, the key feature is that the spray does not contain a muon. The presence of a muon indicates a charged-current interaction, in which the muon-neutrino has become a muon.

However, a high-energy pion can look like a muon in a bubble chamber. To distinguish the two kinds of particle it is necessary to follow their tracks a long way, to discover if the particle in question decays as a pion would, or

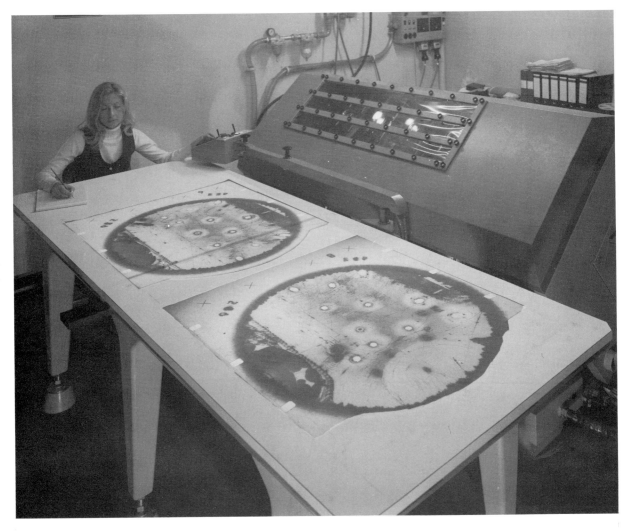

Film from Gargamelle was projected onto tables like this, specially designed for measuring the tracks relative to reference points on the table. (CERN.)

continues, probably to leave the chamber, as a high-energy muon would. Here Gargamelle's length proved crucial. Moreover, neutrons can collide with nucleons to produce muon-less reactions, and neutrons, like neutrinos, are invisible until they interact with something. So neutrons, produced by neutrinos in material *outside* the detector, could enter the detector and mimic neutral-current interactions.

The physicists could reject reactions due to low-energy neutrons – the bulk of the neutrons – by rejecting events in which the total energy measured for the spray of particles was relatively low. That left the high-energy neutrons. The team could only estimate the number of these, but they did have something on which to base their estimates. Sometimes a neutron produced in a spray of particles *inside* the chamber would interact with a nucleus before leaving the liquid, producing a second spray of tracks. By counting the number of times such associated sprays occurred, the researchers could calculate the likelihood of producing a neutron inside the chamber. They could then extrapolate to estimate how many high-energy neutrons might have entered the chamber after being produced outside.

Throughout 1972, a few members of the Gargamelle team worked long and hard in their effort to pin down the elusive neutral currents. Certainly they found photographs with sprays of tracks showing a proton accompanied by pions, but no muon. But were these evidence for neutral currents? Could the team be sure that neutrinos had induced these events, and not the troublesome stray neutrons? At conferences, members of the group such as Paul Musset and Antonio Pullia concentrated on explaining how they were trying to come to terms with the 'background' of unwanted neutron events.

Then, early in January 1973, one of the scanners at Aachen, noticed an unusual event, which she classified as a muon and a gamma-ray. When Franz Hasert, a graduate student, went to look at the picture he realised that the scanner had been mistaken. The tracks, which looked like shepherd's crooks, indicated the path of a lone electron, radiating photons that themselves produced electron–positron pairs in the heavy Freon liquid of Gargamelle (see picture on p.119). What is more, the event had been produced in a beam of antineutrinos, so the chances that the electron came from inverse beta-decay, with the proton unseen, were very small.

The discovery had an important psychological effect. Now Gargamelle had provided indications of neutral currents not only in neutrino–nucleon scattering, but also in neutrino–electron scattering, and many more members of the team joined in the final drive to understand what they were really seeing. Much of the effort, discussion and argument hinged on

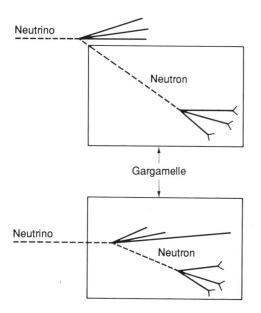

Neutrons produced in interactions just outside the body of Gargamelle could enter the chamber and interact in the liquid to fake the appearance of neutral-current interactions due to neutrinos. To allow for this, the physicists had to count how many times neutrons were made and interacted inside the chamber, in 'associated events'; they could then calculate how many times neutrons probably produced fake neutral-current events.

This image from Gargamelle shows a neutral-current interaction involving a nucleus in the chamber liquid. A neutrino has entered from the left and interacted with a neutron. There is no sign that a muon has been produced – the long track which comes to a halt some distance from the initial interaction point is due to a proton. (CERN.)

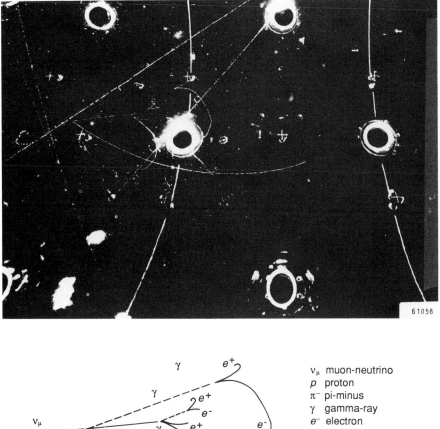

61056

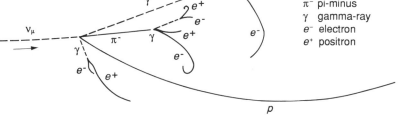

ν_μ muon-neutrino
p proton
π^- pi-minus
γ gamma-ray
e^- electron
e^+ positron

understanding the background of neutron events, as ideas sprang from each of the various centres at which the team was based.

By mid-summer the group had by and large convinced themselves that they were indeed observing neutral currents. In their search for neutrino–nucleon scattering, they had scanned some 83 000 photographs in which neutrinos had interacted, and 207 000 photographs of antineutrino interactions. Of these events, 102 looked like a neutral-current interaction between a neutrino and a nucleus; 64 looked like a similar antineutrino interaction. The likely numbers of neutron-induced events were low – 12 for the neutrino events and 14 for the antineutrino interactions.

The team sent a paper describing the single-electron event to the journal *Physics Letters*, where it arrived on 2 July. A paper describing the neutrino–

nucleon events followed later, arriving on 23 July. Meanwhile, Musset had announced the discoveries at a seminar at CERN on 19 July.

Then followed a series of presentations at different conferences, as the Gargamelle team began the task of convincing particle physicists world-wide of their discovery. Early in September, at Aix-en-Provence, Abdus Salam could at last hear the news first hand:

I remember Paul Matthews and I getting off the train at Aix . . . and foolishly deciding to walk with our heavy luggage to the student hostel where we were billeted. A car drove from behind us, stopped, and the driver leaned out. This was Musset . . . He peered out of the window and said, 'Are you Salam?' I said, 'Yes.' He said, 'Get into the car. I have news for you. We have found neutral currents.' I will not say whether I was more relieved to be given a lift . . . or for the discovery of neutral currents.[12]

It should have been a time of glory for CERN and the physicists working with Gargamelle; but the end of 1973 turned into a nightmare. The team had decided to publish their results partly because they had heard that a neutrino experiment in the US also had evidence for neutral currents. But as the year drew to a close, new results from the US began to cast black clouds over the whole issue.

Even as the team on Gargamelle had been analysing their first pictures, a new breed of high-energy neutrino experiment had begun to take shape in the US. In 1972 a large proton accelerator started up at the National

Paul Musset (right), one of the leading figures in the Gargamelle team, celebrating with Abdus Salam (left) at the time of the announcement of Salam's share in the 1979 Nobel Prize for physics, awarded for his work on the theory that unifies the electromagnetic and weak forces. One of the most important facets of the theory is the prediction of the existence of neutral currents, beautifully confirmed by the experiments with Gargamelle. (CERN.)

Accelerator Laboratory (now Fermilab), on the Illinois plain to the west of Chicago. This machine accelerated protons to energies of up to 400 GeV, more than 15 times the energy of the protons from the PS at CERN. Some distance from the accelerator ring, at one end of a mound of earth 1 kilometre long, lay the Neutrino Area. The mound served to filter out all unwanted particles in the neutrino beam created in the decays of particles produced by the high-energy protons from the accelerator.

The first neutrino experiment at Fermilab involved a two-part detector, put together by a team from the Universities of Harvard, Pennsylvania and Wisconsin and from Fermilab. The first part, as seen by the muon beam, was a 'target-detector' – a series of 16 large flat containers filled with liquid scintillator, interspersed with spark chambers. When neutrinos interacted in the scintillator, they would create a spray of charged particles. Phototubes would pick up all the light produced as the particles passed through the scintillator, and give a measure of the total energy of the spray; the spark chambers would reveal the tracks of the particles.

The second part of the detector consisted of four large 1-metre thick pieces of magnetised iron, again interspersed with spark chambers to pick up particle tracks. This section was intended to detect muons – the only particles likely to penetrate far through the iron – and it could measure the momentum of the muons from the bending of their tracks in the magnetic field. Overall, the detector contained 70 tonnes of liquid scintillator, in containers 4 metres square. This made it more likely to intercept neutrinos than Gargamelle, which had only 10 (visible) tonnes of liquid.

In the summer of 1973, the first results from the experiment at Fermilab apparently confirmed Gargamelle's discovery of the muon-less events characteristic of neutral currents. In August, the team in the US sent a draft paper to *Physical Review Letters*, but it was destined to remain unpublished until the following April, for there followed a troubling autumn.

When the team at Fermilab looked for muon-less events in a supposedly improved version of their detector, the number of such events seemed greatly reduced. There now appeared to be a direct conflict between the results from Fermilab and those from Gargamelle. A state of near-panic struck some people in the upper echelons at CERN, with the thought that the published results from Gargamelle could be wrong; but the team managed to stand firm. The group at Fermilab, meanwhile, had by mid-November drafted a new paper claiming that they had no evidence at all for neutral currents. However, it was never submitted, for as they continued their attempts to come to terms with understanding their apparatus, they were led gradually to the conclusion that muon-less events, and hence neutral currents, did exist after all!

It was only in February of the following year that the team on the Fermilab experiment had ironed out all their problems, and at last felt confident enough to publish their original paper on 'Observation of Muonless Neutrino-Induced Inelastic Interactions'. By April a third neutrino experiment, this time in a 4-metre bubble chamber at the Argonne

(a) This aerial view of Fermilab in 1976 shows the mound of earth covering the proton beam as it leaves the main accelerator ring (visible in part to the right, marked by the service road that follows the ring round on the surface). The proton beam is later split at the 'switchyard' to serve different facilities.

One of the beams it generates is the neutrino beam which continues into the distance before reaching the experiments in the neutrino area, shown in a closer view in (b). (Fermilab.)

Laboratory, in Illinois, had also found conclusive evidence for neutral currents.

'Do neutral currents exist or not?' asked André Rousset at the 'Neutrino '74' meeting in Pennsylvania in April 1974. He hedged his bets by concluding that

the neutral currents appear today to theoreticians as the most suitable interpretation of the results obtained by the experimentalists.[13]

Two months later, Don Perkins was more assertive when he said of neutral currents:

The effects have been observed in four independent experiments at three accelerators, and their existence is firmly established . . . The discovery . . . at the levels observed, is very strongly suggestive, for the first time, of a basic unification of two of [the fundamental] interactions, the weak and electromagnetic. This is a profound step forward.[14]

From this point on, the study of neutral currents became a topic in its own right. As many people were quick to point out, the discovery of neutral currents did not in itself prove that the unification of Weinberg and Salam was correct; other kinds of theory could also require neutral currents. The time had come to study the neutral currents and to confirm that they did indeed behave in the way that the electroweak theory predicted. Meanwhile, the Gargamelle experiment was enriching knowledge of a different aspect of subnuclear behaviour, leading physicists deep within the proton.

Inside the proton

> . . . it's very much like the days of Rutherford . . . by scattering these things on a proton, we have found that although at first we might have thought that the charges were somehow smeared around in that space, and nature was smeared in space, it turns out no, nature has . . . sharp point-like insides.[15]
>
> *Richard Feynman, April 1974.*

In the same year that Lagarrigue and his colleagues set out their proposal for Gargamelle, theorists Murray Gell-Mann and George Zweig put forward their ideas for a new layer to the structure of matter: the quarks. But many particle physicists were slow to accept the notion of quarks. After all, these proposed constituents of protons and neutrons had some bizarre properties, in particular fractional values of $\frac{1}{3}$ and $\frac{2}{3}$ the unit of electric charge on an electron or proton. By 1971, however, when the Freon-filled giantess finally heaved, shook and began to swallow her first neutrinos, the perception of quarks was subtly changing. Experiments were beginning to reveal that there might be something more to quarks than a tantalisingly beautiful mathematical symmetry.

The key experiments had taken place in California at SLAC, the Stanford Linear Accelerator Laboratory. There, in the summer of 1967, a team of researchers from SLAC and the Massachusetts Institute of Technology

(MIT) began to use a powerful new instrument to probe the interior of the proton. An accelerator 3 kilometres long shot electrons at energies of up to 16 GeV into a target of liquid hydrogen. Banks of detectors picked up the electrons as they ricocheted off at different angles, having lost varying amounts of energy in their collisions with protons in the target.

One area of interest concerned the 'inelastic' collisions – interactions in which new particles are created. Here it was soon clear that something unexpected was happening. From their measurements, the physicists could calculate the momentum transferred from the electron to the proton in each collision they detected. And they found that the probability for reactions was surprisingly high when larger amounts of momentum were transferred. Indeed, the probability for the high-energy inelastic reactions seemed hardly to vary at all with momentum transfer when compared with the theory for simple scattering from a point of electric charge. In other words, high-energy inelastic scattering appeared very similar to 'point-like' scattering. This was in marked contrast both to elastic scattering and to inelastic scattering at lower energies. In these cases, the relative probability dropped rapidly with increasing momentum transfer.

What was happening? As far as we can tell, electrons are like simple points with no size; they do not behave as if they are spread over some region in space. We know this because the theory of quantum electrodynamics works so well in describing the behaviour of electrons, and this theory treats electrons only as simple points. By contrast, experiments that scatter low-energy electrons from protons show that protons do have some size. Their electric charge is spread over a region in space. Because of this 'smearing', low-energy electrons can pass *through* protons with relatively little deviation.

But the experiments with high-energy electrons at SLAC were revealing a different picture of the proton. Here the electrons could give up large amounts of momentum to the proton, and be scattered through unexpectedly large angles. This could not happen if the electron was 'seeing' a smeared-out proton; it could happen only if the electron was coming up against some small, point-like concentration of electric charge within the proton. It was in a sense a rerun of the story of Rutherford's discovery of the atomic nucleus, only this time high-energy electrons replaced alpha-particles as the missiles, and they were being knocked sideways, not by the nucleus, but by something hard *within* the nucleus.

In the summer of 1968, at the Fourteenth International Conference on High-Energy Physics in Vienna, Wolfgang Panofsky from SLAC presented a review talk on electron scattering. Referring to the first results on high-energy inelastic scattering from the SLAC–MIT group he commented:

The qualitatively striking fact is that these cross-sections . . . decrease much more slowly with momentum transfer than the elastic scattering cross-sections . . . Therefore theoretical speculations are focused on the possibility that these data might give evidence on the behaviour of point-like, charged structures within the nucleon.[16]

During the early 1970s, the Stanford Linear Accelerator Center (SLAC) produced compelling experimental evidence for structure in the proton. The centre houses an electron linac (linear accelerator), extending 3 kilometres from the electron source, at the bottom of this picture, to the area where the experiments are located on the other side of the freeway. (SLAC.)

Among those speculating was James Bjorken, a theorist at SLAC. In lectures he had given in July 1967 at the 'Enrico Fermi' International School of Physics in Varenna, Bjorken had shown that when extrapolated to high energies, theoretical expressions describing electron–proton scattering resemble those for scattering from point charges. He believed that the relations he found were

. . . so perspicuous that, by an appeal to history, an interpretation in terms of 'elementary constituents' of the nucleon is suggested.[17]

In a later paper, submitted to *The Physical Review* at the end of September 1968, Bjorken also showed how at high energies the process of inelastic electron scattering should exhibit a property that became known as 'scaling'.

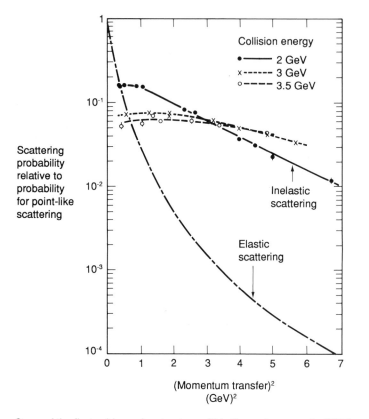

Some of the first evidence for structure within the proton came in 1969 from the new linac at SLAC, with studies of *inelastic* interactions between electrons and protons, in which additional particles are created. The measurements showed that the electrons had a surprisingly high probability of transferring large amounts of momentum to the protons, just as if they were scattering from simple points. Here the probability measured by the SLAC–MIT experiment has been divided by what is expected for pure point-like scattering, and can be seen to vary very little with transferred momentum. Results for *elastic* scattering (no additional particles created) show a very different variation. (*Physical Review Letters* **23**, 935, 1969.)

Richard Feynman lecturing in 1965, the year he shared the Nobel Prize for his work on the quantum theory of the electromagnetic force. Three years later, he began to use his 'parton model' of protons – in which 'partons' within a proton share its momentum – to understand the exciting new results from SLAC on inelastic electron–proton scattering. (CERN.)

He predicted that the results would show no dependence on any dimension, but would depend only on a dimensionless ratio derived from the energy and the momentum transferred between electron and proton. In other words, the reaction would reveal no characteristic 'scale'.

One visitor to SLAC in the summer of 1968 was Richard Feynman, the charismatic and brilliant theorist who had already shared the Nobel prize in 1965 for his work on the development of quantum electrodynamics. In 1968 he was working on what he called the *parton model* of protons, in which he considered the proton to consist of a number of pieces – *partons* – each of which carried a certain fraction of a proton's momentum. As Feynman explained:

The quantities that characterise our proton are then distributions like the statistical probability that a particular kind of part carries a fraction x of the proton's momentum.[18]

Feynman had developed his model in the context of high-energy collisions between *hadrons* – strongly interacting particles, such as pions, kaons and protons. He had been disappointed with his progress, but when he learned of the new results at SLAC and Bjorken's ideas on scaling, he immediately recognised their significance:

I saw that the experiments were tailor-made for investigating partons, and were easy to interpret in terms of the pictures I had already developed for the strong interactions. [These] experiments can identify the kinds of partons and how they are distributed . . .[19]

According to Feynman, the simplest way of looking at the inelastic electron-scattering experiments was to transpose them to a different reference frame (a ploy that particle physicists often use), in which the proton and electron are moving towards each other, close to the speed of light. The proton would then appear as a parallel stream of partons, and:

For a collision at such [high] momenta (where energy and momentum are practically equal) the conservation laws of energy and momentum simply say that the particles exchange momentum. The momentum of the electron scattered back gives directly the momentum of the part from which it scattered . . . Thus the momentum distribution of the returning backscattered electron gives directly the distribution of the *charged* parts . . . the electron would not scatter from neutral parts.[20]

In other words, measuring the momentum of the scattered electrons provides a mirror on the way the parts inside the protons share the protons' momentum.

Feynman possessed an insight into fundamental processes that was the envy of many physicists, and which seemed to make him able to come straight to the point and describe clearly what was happening. He likened the electron scattering experiments to studying a swarm of bees with radar:

If the swarm is moving as a whole the frequency of waves reflected back determines its speed. But if instead individual bees are moving about in the swarm, the returning wave has a range of frequencies corresponding to the range of velocities of the bees in the swarm.[21]

By the same token, the electrons scattered by the protons have a range of momenta corresponding to the range of momenta of the parts in the proton.

One of the most important feature of the parton model was that it provided a physical explanation for Bjorken's dimensionless ratio. It turned out that this was the same as x in Feynman's parton model – the fraction of the proton's momentum carried by a given parton. Equivalently, for a proton at rest in the laboratory, rather than in the high-energy reference frame, x is the fraction of the proton's mass carried by the parton.

Feynman had responded quickly to the discoveries at SLAC, and Bjorken was equally quick to respond to Feynman's ideas. Working with Emannuel ('Manny') Paschos, Bjorken soon incorporated the concept of partons into a now classic paper, published in 1968. But as Jack Steinberger later remarked, 'For mortals the appreciation came more gradually.'[22]

The first results on inelastic scattering from SLAC, which Panofsky had presented at Vienna in 1968, were at only one angle. By 1970, however, at the Fifteenth International Conference in Kiev, the SLAC–MIT team had measurements at six different angles. These new data demonstrated 'Bjorken scaling' over a wide range of transferred momentum. 'Now,' recalls

Steinberger, 'the fact that the proton contains quasi-free point-like constituents was established and accepted.'[23]

Partons as quarks

> The miraculous success of the parton description [of the proton] naturally leads to two questions: what is the nature of the partons . . . and why does such a simple model work so well?[24]
>
> *Don Perkins,* Introduction to High Energy Physics.

> Taken together, the data also point to the strong possibility that the partons are quarks, though this interpretation does raise some problems because free quarks have not yet been seen . . .[25]
>
> *Peter Landshoff, May 1974.*

So Feynman's partons apparently existed, but were the partons indeed the same beasts as the quarks of Gell-Mann and Zweig? Particle physicists were careful not to jump to the obvious conclusion too readily. The answers would eventually be found by experiment; the important point was to do the right experiments and to interpret the data correctly.

The first encouraging sign came from the electron experiments at SLAC, which showed that the partons must have a spin of $\frac{1}{2}$ as was the case for the quarks. But what about the puzzling fractional charge of the quarks? According to the quark picture, the proton contains two positive u quarks with charge $\frac{2}{3}$ the size of the electron's charge (e), and one negative d quark with a charge of $-\frac{1}{3}e$. Was there a way to measure the electric charges of the partons, and to discover if they too carried charges of $\frac{1}{3}e$ and $\frac{2}{3}e$? Indeed there was. The answer lay in comparing data from the electron-scattering experiments with the results of delving inside the proton with a different probe – the neutrino, which like the electron, behaves as a simple point, with no smeared-out structure.

The giant bubble chamber Gargamelle began to observe the very rare interactions of muon-neutrinos in 1971, just as the full significance of the results from SLAC was beginning to take hold. A neutrino passing through Gargamelle would from time to time induce a spray of curling tracks due to a bunch of charged particles. Often one of the tracks would be straighter than the others, indicating the passage of a lightweight, energetic muon. These 'charged-current' events, in which the incident muon-neutrino scattered off a nucleon and changed into a muon, were equivalent to the inelastic electron-scattering detected at SLAC.

Electrons interact with protons through the electromagnetic force – in quantum terms, through the exchange of a photon. Neutrinos, however, interact only weakly, most frequently exchanging a charged W particle to give the muon-producing, charged-current events. When a muon-neutrino interacted inelastically with a nucleon in Gargamelle, it would change into a negative muon, emitting a positively charged W (W^+) to conserve charge. The nucleon would absorb the W^+, become 'excited', and create several particles as it returned to a normal state. A muon-antineutrino would in a similar way emit a W^-, and change into a positive antimuon.

But what was happening at the level of quarks? In the 'quark–parton' model, the two u quarks and one d quark that the proton contains are called 'valence' quarks, in analogy to chemical terminology, where the valence electrons are the outer ones that determine the properties of an element. This name is to distinguish the three quarks that govern the properties of the particle they form from an ephemeral cloud or 'sea' of quark–antiquark pairs in which they are embedded according to the quark–parton picture. These pairs are continually forming from the field presumed, in the parton theory, to exist within the proton, rather as electron–positron pairs form in an electromagnetic field; and, just as electrons and positrons annihilate, the quarks and antiquarks readily recombine.

In this picture, the probability for neutrino scattering from a proton depends on the momentum carried by the d quarks in the proton and also upon any u antiquarks it contains. This is because when a neutrino changes to a muon, it can emit only a W^+ to conserve charge, and the W^+, with its one unit of positive charge, can only change a d quark (charge $-\frac{1}{3}e$) into a u quark (charge $+\frac{2}{3}e$), or a u antiquark (charge $-\frac{2}{3}e$) into a d antiquark (charge $+\frac{1}{3}e$). Similarly, the probability for antineutrino scattering, in which a W^- changes hands, depends on the u quarks and d antiquarks. So summing the probabilities for both neutrino and antineutrino scattering adds together the contributions from all quarks (u and d) and antiquarks ($\bar{u}$ and $\bar{d}$) in the proton. A similar line of arguing applies to scattering from neutrons.

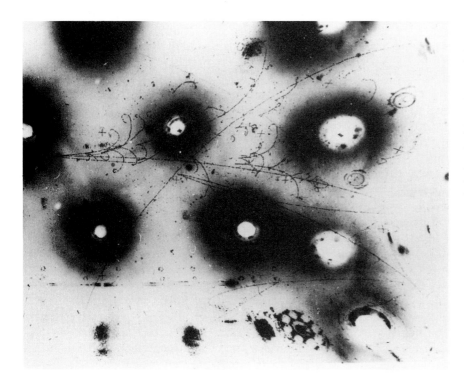

A typical inelastic charged-current interaction in Gargamelle in which a neutrino (coming in from the left) exchanges a W particle with a proton or neutron in a nucleus in the liquid, and emerges as a muon, producing a long track that leaves this picture on the right as the third track from the bottom. Information culled from many events of this kind was vital in establishing that the partons – 'parts' within protons and neutrons – are indeed fractionally charged quarks. (CERN/Courtesy G. Myatt.)

When an electron scatters inelastically from a proton it transfers energy and momentum to the proton via a photon, and the energy materialises as several new particles. A similar process occurs in inelastic neutrino scattering, but in this case a *W* particle transfers the energy and momentum to the proton.

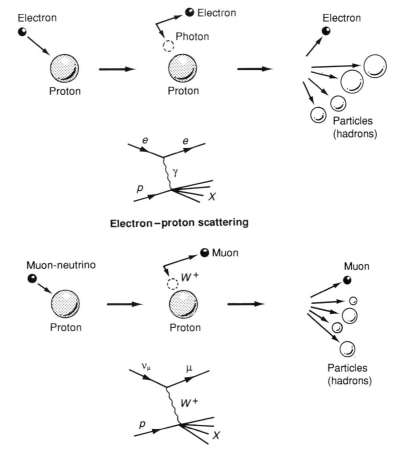

Electron–proton scattering

Neutrino–proton scattering

Gargamelle contained Freon, with neutrons as well as protons, so the physicists could measure only the *average* scattering from protons and neutrons. But this was sufficient for some important discoveries. First, the shape of the momentum distribution was the same as that determined in a similar way from the electron-scattering data. However, a comparison of the two sets of results led a step farther. This is because electrons 'see' the charge on the quarks, and the probability for scattering is proportional to the square of the charge − $\frac{4}{9}$ for the u quarks and $\frac{1}{9}$ for the d quarks. As Feynman commented,

> If you never did believe that 'nonsense' that quarks have nonintegral charges, we have a chance now, in comparing neutrino to electron scattering, to finally discover for the first time whether the idea that quarks have nonintegral charges is physically sensible, physically sound . . .[26]

adding in masterly understatement, 'that's exciting'.

The effect of the charges of the different quarks brings a factor $\frac{1}{2}(\frac{4}{9}+\frac{1}{9})=\frac{5}{18}$

into the probability for the average electron scattering from protons and neutrons. So if the momentum distribution from Gargamelle was multiplied by $\frac{5}{18}$, did it reproduce the momentum distribution from SLAC?

'The gross cross-sections agree perfectly with the predictions of the fractionally-charged quark model',[27] Don Perkins, a leading member of the Gargamelle team, told physicists (including Feynman) when he presented preliminary measurements from Gargamelle, at a meeting in Hawaii in August 1973. He later commented:

The evidence is rather compelling that electrons and neutrinos are seeing the same substructure inside the nucleon, with absolute rates standing in exactly the ratio predicted by the quark charge assignments.[28]

So, there *are* fractionally charged objects within protons and neutrons. But the original quark model of Gell-Mann and Zweig assigned three quarks each to the proton and the neutron. In the quark–parton picture, these are the valence quarks, *uud* in a proton and *udd* in a neutron. Can we check this assignment with neutrinos? Perhaps surprisingly, the answer is 'yes'.

In this case the test is to subtract the measured probability distribution for antineutrinos from that for neutrinos. This gives the difference between the momentum distributions for the quarks and the antiquarks – in other words, it should depend on the total number of quarks q in a nucleon minus the number of antiquarks, $\bar{q}$. Any 'sea' of quark–antiquark pairs should give equal numbers of q and $\bar{q}$, so the difference $q - \bar{q}$ should reveal the number of

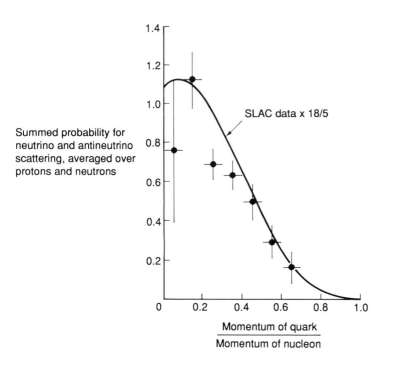

Summed probability for neutrino and antineutrino scattering, averaged over protons and neutrons

SLAC data x 18/5

Momentum of quark
——————————
Momentum of nucleon

A comparison of measurements from electron scattering at SLAC with neutrino scattering in Gargamelle not only shows that the two different types of particle 'see' the same sharing of momentum between the parts inside the nucleons; by observing that the results for neutrino scattering follow the same curve as those for electron scattering multiplied by $\frac{18}{5}$, we have good evidence that the parts are indeed quarks with charges of $+\frac{2}{3}$ and $-\frac{1}{3}$ the charge on the proton. (*Nuclear Physics* **B85**, 269, 1975.)

These measurements from Gargamelle confirm that quarks rather than antiquarks carry momentum in a proton or neutron. This reinforces the picture in which a proton or neutron consists of three 'valence' quarks, in an ephemeral 'sea' of quark–antiquark pairs appearing and disappearing from the field of the strong force that binds the valence quarks together. The curves are from various theoretical predictions. (*Nuclear Physics* **B85**, 269, 1975.)

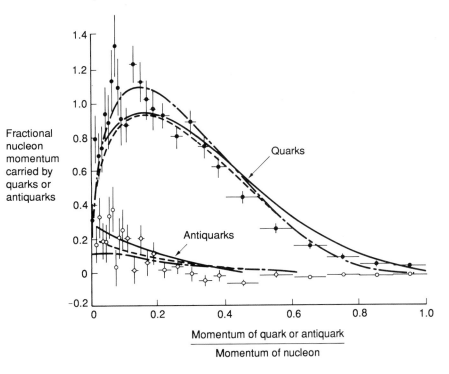

valence quarks. Once again, the preliminary results from Gargamelle supported the quark model, by indicating that $q - \bar{q}$ was indeed three.

A third test using the results of the neutrino experiments was to check the total fraction of a nucleon's momentum that is carried by all the quarks and antiquarks. This information comes again from the combined measurements for both neutrinos and antineutrinos (and can also be extracted from the data for inelastic electron scattering). If there are only quarks and antiquarks within a nucleon, then adding up the fractions of momentum they carry should give a total of 1, the whole momentum for the nucleon. Instead, the results add up to about 0.5. It seems that only half the momentum is carried by quarks and antiquarks. There must be something else inside the proton and neutron. And this something else must interact neither with electrons nor with neutrinos; it must feel neither the electromagnetic force nor the weak force.

This result was not altogether surprising. After all, something must hold the quarks together in a nucleon. Indeed, as Feynman commented:

It might, in fact, have been embarrassing if all the momentum were accounted for by quarks. We might not know how to describe their interaction.[29]

Particle physicists knew that some nuclear interactions, like those that bind protons and neutrons together in nuclei, are not only far stronger than the effects of the weak force, but much stronger than the electromagnetic force between charged particles. What else can hold a nucleus together? Why is it not blown apart by the repulsive forces between the like-charged protons? It seemed likely that whatever was holding the quarks together was probably at the root of this strong nuclear force.

According to quantum field theory, the force between quarks must be carried by a field particle, just as the electromagnetic field is carried by photons and the weak force by W and Z particles. In the case of the interquark force someone had coined the name of 'gluon'. The results from SLAC and Gargamelle hinted for the first time at the real existence of gluons, darting between the quarks within a nucleon.

In the giant's footsteps – an epilogue

> . . . nothing is ever a given moment in history, we are always making steps . . .[30]
> *Richard Feynman, April 1974*

The Gargamelle experiment was in a sense a bridge. It linked the 'old' physics of the 1960s to the 'new' physics of the 1970s. It led particle physics from a world inhabited by many kinds of subatomic particle, held together by distinct forces, towards a promised land of simple subatomic building blocks united by a single underlying force. Hindsight allows us to focus on this one experiment, for we know now that the small handfuls of neutrino scatters that Gargamelle gathered gave a true sampling of the nature of things. At the time, it was not always so clear.

Experiments with neutrino beams are plagued with difficulties. Not only can you not 'see' the beam, but any interactions that do take place are extremely rare and in danger of being swamped totally by unwanted 'background' effects. With Gargamelle, as with every neutrino experiment before and since, physicists have argued about the measurements – about the calculations of background effects, about the energy and intensity of the invisible neutrino beam, about the responses of the different detectors. Such discussions occur over any particle physics experiment, but with neutrinos the arguments are more difficult to lay to rest and seem unusually capable of exciting passionate debates between stubborn protagonists.

The story of neutral currents, with the 'now you see them, now you don't' episode at Fermilab, provides a prime example of the arguments that can arise with neutrinos. But it also illustrates how greatly rewarding discoveries do come from what can appear the most frustrating of experiments. Since the first results from Gargamelle and Fermilab, a number of experiments, mainly at CERN and Fermilab, have performed many more detailed studies of neutral currents. A large number of results have come from CERN's Super Proton Synchrotron (SPS), which started up in 1976. Like the original machine at Fermilab, the SPS accelerates protons to energies of 450 GeV, and these can be used to create a beam of high-energy neutrinos. Over the years, several detectors there – including Gargamelle, after being moved from the lower energy PS in 1976 – have collected a mass of data on neutrino interactions.

By the end of the 1970s, these experiments had come down firmly in

In the late 1970s and during the 1980s, several new experiments at CERN took the studies of neutrino scattering initiated with Ramm's chamber and Gargamelle to higher energies still. This experiment, CHARM II, started up in 1986, its main task to make precision measurements of the rare neutrino–electron scattering. The section at the rear of this view contains 692 tonnes of glass as a 'target' for the neutrinos, interspersed with particle detectors; the section in the foreground is to detect and measure muons produced by the muon-neutrinos. (CERN.)

support of an electroweak model of the kind proposed by Salam and Weinberg. In 1979, Weinberg and Salam shared the Nobel prize for physics, together with Sheldon Glashow, who had been instrumental in showing how to incorporate the weak interactions of quarks into electroweak theory. (Glashow's work had led him to discover in 1971 the necessity for the charm quark, c, which was at the time unknown.) Later experiments led to still more precise tests of the electroweak theory, proving how well it works down to distances of 10^{-16} cm – or one thousandth the distance across a proton.

As far as the quark–parton discoveries were concerned, while the early results from Gargamelle had suggested that the approach was correct, there was at the time no underlying quantum field theory, akin to the electroweak theory, to describe what was happening deep in the heart of a proton. Indeed, the very success of the ideas was paradoxical. The parton description assumed that the incoming electron or neutrino interacted with an individual part of the proton, and ignored the other parts. But the quarks were supposed to interact strongly via the exchange of gluons. So if partons were quarks and gluons, why did interactions between them not influence the 'scaling' effects? It seemed as though when you probed deeply, at high transferred momentum, the partons appeared only weakly bound. But what kind of force could at the same time become so strong that it prevented single quarks from being ejected from a nucleon?

The apparent weakness of the force at high transferred momentum – or equivalently at short distances within the proton and neutron – was an important clue. As early as 1973, several theorists realised that such an effect occurred naturally in a certain class of field theories, and this led over the next few years to the development of *quantum chromodynamics*, or *QCD* – the quantum field theory of the strong force.

In QCD the strong force between quarks originates in a property called *colour*. Here the use of the word colour bears no relation to its normal meaning; instead it refers to a property of quarks that is in a sense analogous to electric charge. Just as electric charge gives rise to the electric field around charged particles, so colour gives rise to a 'colour field', or in other words, the field of the strong force.

The name colour arises because, like the primary colours of light, this property of quarks appears to exist in three varieties. This is in contrast to electric charge, which appears in only one form – the charges we call positive and negative being in effect 'charge' and 'anticharge'. And rather as an atom is electrically neutral overall, although it consists of negative electrons and a positive nucleus, so particles that contain quarks must be 'colour neutral' overall. A proton, made of three quarks, must contain one quark of each colour, so that the colours add together to give zero colour; an *anti*proton contains three antiquarks carrying three *anti*colours. (The idea of colour arose originally from attempts to explain why certain particles can contain three otherwise identical quarks, in apparent violation of a fundamental principle of quantum theory – the Pauli principle.)

Because there are three varieties of colour, it turns out that there must be as many as eight different kinds of gluon – the ball in this quantum game of 'strong catch' – and that the gluons themselves carry colour. (Strictly speaking they carry a combination of colour and anticolour.) This is quite different from the photons in quantum electrodynamics, which are electrically neutral, and it leads to the radically different behaviour of the strong force. Not only can gluons change the colours of quarks they pass between, but they can also act upon each other. A useful comparison is with a beam light, in which photons stream out independently across space. A similar beam of gluons would not travel very far because the gluons would interact between themselves.

This behaviour of gluons helps to explain both why quarks cannot escape alone from inside nucleons, and also why the quarks within a nucleon appear uninfluenced by each other when probed at short distances with high momentum. Between each quark must lie an entangled swarm of gluons, which, in effect, shields the quarks from each other. This apparent 'freedom' of quarks at high values of momentum was a blessing for the theorists. It meant that to predict the outcome of experiments, they could use the same calculational techniques for the strong force as they used for the more feeble electromagnetic and weak forces. But at low momentum they were left with the problem of being unable to calculate the effects of the undiluted strong force.

Can the influence of the gluons be seen in the inelastic electron and neutrino scattering experiments? The answer is 'yes', because they can modify the way that momentum is shared among the constituent parts of a nucleon. A gluon can 'split' into an quark and an antiquark and the probing electron or neutrino can scatter from one of these two particles; or a quark can radiate a gluon and lose momentum before scattering the probing

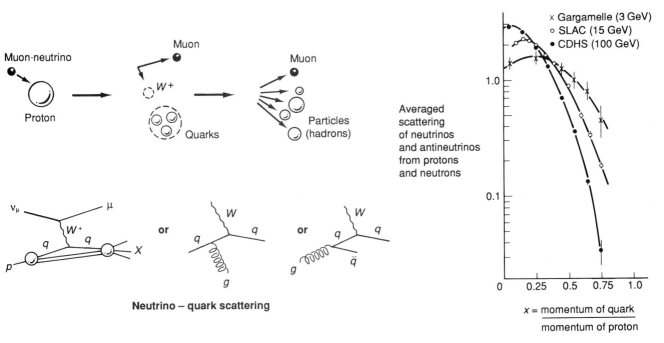

Neutrino – quark scattering

Averaged scattering of neutrinos and antineutrinos from protons and neutrons

x Gargamelle (3 GeV)
o SLAC (15 GeV)
● CDHS (100 GeV)

$x = \dfrac{\text{momentum of quark}}{\text{momentum of proton}}$

At higher energies, gluons – the carriers of the strong force – are more likely to modify the neutrino's view of the way that the quarks within a proton share its momentum. A quark that radiates a gluon just before it interacts with a neutrino by absorbing a W particle, will have a smaller momentum than otherwise. Moreover, at high energies a gluon has more chance of producing a quark–antiquark pair, each with half the momentum of the gluon. Both these effects combine to squeeze the momentum of the quarks to lower values – easily seen in this comparison of results at increasing energy from Gargamelle, SLAC (with electrons), and from the CERN–Dortmund–Heidelberg–Saclay experiment (CDHS).

particle. Processes such as these should become more noticeable as the proton is probed more closely – in other words, in the scattering where larger amounts of momentum are transferred. The additional interactions will tend to reduce the average momentum of the partons, and should modify the measured momentum distributions, so that 'scaling' no longer holds so well – in other words, the distributions should no longer depend only on the simple dimensionless ratio, x, but should vary as the energy increases, probing the proton on smaller scales.

The first evidence for the modifying influence of the gluons came in an experiment at Fermilab that used high-energy muons, rather than electrons or neutrinos, to probe the nucleon. In general, the results were in line with those from the electron and neutrino experiments, showing no dependence on the momentum transferred between the muon and the nucleon. However, at high values of momentum transfer there were small but significant deviations from the usual scaling effect.

Measurements with high-energy neutrino beams show the same slow deviation from scaling at high values of momentum transfer. By fitting the predictions of quantum chromodynamics to the data from both neutrino and muon experiments, researchers have been able to discover the momentum distribution of the gluons. The gluons turn out to carry small fractions of the momentum of the nucleon – small values of x – but as there are many of them, together they carry as much as half the total momentum.

These results on gluons are only part of a broad picture of the inside of the proton that the neutrino experiments have helped to expose. In particular, the 'handedness', or helicity, of the neutrino has imbued it with a special role, for it means that neutrinos interact differently with quarks and antiquarks. This has allowed researchers to extract the momentum distribution for the antiquarks in the quark–antiquark 'sea' inside the nucleon. Like the gluons, these antiquarks carry small fractions of the overall momentum.

All in all, neutrino 'spaceships' have provided some of the most beautiful evidence for the point-like structure inside protons, over a wide range of neutrino energies – from 2 to 200 GeV. In Perkins' words:

> Perhaps the most dramatic demonstration of the constituent nature of hadrons is provided by the total cross-section, as a function of energy for neutrino–nucleon scattering . . . the result is a simple one – an almost linear rise of [the cross-section] with the neutrino energy. This is exactly the result we expect if we replace the complicated process of hadron production . . . by the *elastic* scattering of the neutrino by a *single pointlike particle*.[31]

With pointlike particles, there is no structure to affect the reaction probability – that is, the cross-section – which depends simply on the strength of the weak force, and on the energy available. The neutrino data show clearly how even at the higher energies, the quarks still appear like points. The only change with energy is very slight, and is completely accounted for by the gluons of quantum chromodynamics. (It is interesting to note that results from Colin Ramm's heavy liquid bubble chamber at CERN provided evidence for this simple dependence on energy, several years before the first indications of 'scaling' were seen at SLAC. Yet no one interpreted the neutrino data in terms of the simple picture of nuclear constituents, no doubt because the general belief was that the nucleus was a complex place.)

One additional way in which neutrinos have helped to probe the nature of quarks is in experiments that have studied the production of particles containing the heavier types of quark, *c* and *b*. These quarks did not figure in the original proposals of Gell-Mann and Zweig, but were discovered in new

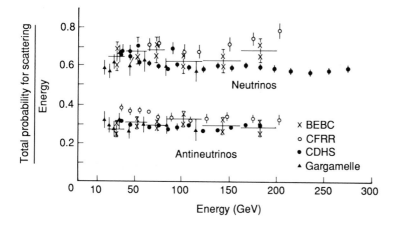

The total probability for neutrino (and antineutrino) scattering from nucleons, when divided by energy remains constant to the highest energies studied so far – in other words, it depends simply on energy. This provides the most dramatic evidence that nucleons contain objects that continue to behave like simple points up to the highest energies investigated. (The data are from four different experiments.)

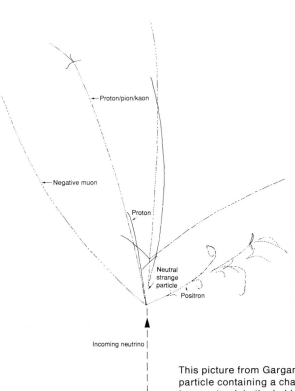

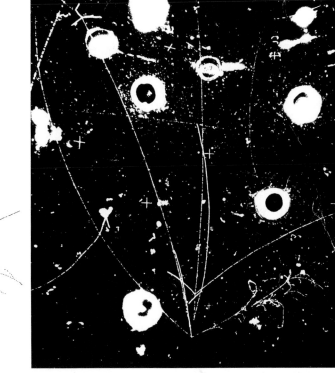

This picture from Gargamelle shows the production and decay of a charmed particle – a particle containing a charm quark. The charmed particle does not live long enough to leave a track in the bubble chamber, but is recognisable through its decay to a positron and a neutral strange particle, which itself decays to two charged particles, producing a pair of tracks that make a characterisstic 'V' shape. (CERN/Courtesy G. Myatt.)

particles created in high-energy experiments during the 1970s. The c and b quarks are both much heavier than the u and d quarks of the nucleon, so the particles they form are correspondingly much more massive. They soon decay to lighter particles, containing lighter quarks, after brief lifetimes in the region of 10^{-12}–10^{-13} seconds. In these decays, the c or b quarks change through the agency of the weak force, to s quarks, in processes analogous to the beta-decay of the neutron.

Experiments of various kinds have studied the production and decays of particles containing c and b quarks, and neutrino experiments have played their own complementary role, making use of the neutrino's unique properties. They have, for example, provided some of the first examples of individual events showing the decays of charmed particles – in other words, particles in which a c or charm quark, is unaccompanied by its antiquark, $\bar{c}$.

As early as 1975, the year after the c quark had been discovered in the particle called the J/psi (in which it is bound to a c), Nick Samios's team at the Brookhaven National Laboratory claimed to have discovered what was possibly a charmed particle, containing a *c* quark together with two other quarks. Their experiment was with a neutrino beam in the 200-centimetre hydrogen bubble chamber. By the following year, both Gargamelle and the 5-metre bubble chamber at Fermilab had several examples of what could be charmed particles.

The days of experiments with neutrino beams are now largely over. Gargamelle is no more, and only one large detector remains in the neutrino beam line at CERN. As particle physicists have turned to higher energies to test the ideas of the electroweak theory and QCD – which together form what is known as the 'Standard Model' – they have turned to different kinds of experiment. In particular, by colliding counter-rotating particle beams in an accelerator ring, they have been able to reach energies high enough to release the long-sought charged *W* particles and their neutral partner, the Z^0

The neutrino's role as a direct probe of matter may be largely past, yet its presence is often a key feature in the 'events' studied at particle colliders. Of course, the neutrinos produced in the head-on collisions leave no tracks. But as in the experiments on beta-decay that led to Pauli's prediction of the neutrino, the neutrino's presence is betrayed by its apparent absence – by the energy that it takes away with it.

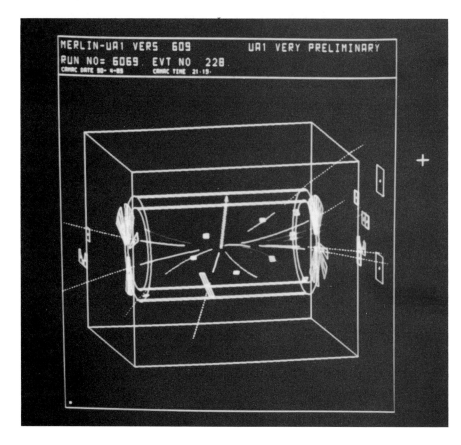

The decay of a *W* particle is captured in this computer reconstruction of tracks in the UA1 detector at CERN. The *W* was created in the annihilation of a proton and an antiproton at the centre of the detector; it has decayed into an electron, the track that points down at a '7 o'clock' position, and a neutrino. The neutrino does not leave a track, but by adding together the energies of everything else produced in the annihilation, and calculating the missing energy, the computer has deduced the path taken by the unseen particle and has drawn in the track pointing upwards that is marked with crosses. (CERN.)

In CERN in 1984, collisions of protons and antiprotons revealed a few lop-sided events, with an energetic muon or electron heading off in one direction, apparently unbalanced by any energy leaving the opposite way. 'Like the sound of one hand clapping', as someone commented at the time, these events signalled the decay of the W particle into an electron or muon accompanied by the appropriate neutrino: the missing energy together with the charged particle's energy added to give the mass of the W particle – in beautiful agreement with the predictions of the electroweak theory.

In similar experiments in the latest colliders at CERN, Fermilab and SLAC, physicists will continue to search for the elusive neutrino's calling card, revealing perhaps new particles or new processes. In these high-energy experiments, the aim is to investigate how matter behaved in the very earliest moments of the Universe. The neutrino will no longer be a spaceship that we send in to probe the nucleus; it will be a spaceship that comes out from that inner space, and will bring with it important messages about the origins of matter.

6

Solar spaceships

The neutrinos provide . . . the only way to penetrate the massive shield of a star's body and see what the centre is like . . . this message comes constantly to us riding a beam as bright . . . as sunlight itself, and yet we cannot detect it at all![1]

Philip Morrison, 1962.

THE SUN is an unremarkable star: 'In the community of stars . . . a respectable middle-class citizen',[2] according to Arthur Eddington. Yet it is vastly interesting. As the nearest star by far, it is the one that we understand the best; moreover, it has the fundamental attraction of being the dynamo that drives life on Earth. The sunlight we see comes only from its surface layers, but the energy that fuels life itself originates far deeper, in a hot, dense thermonuclear furnace buried at the Sun's core

Sunlight takes eight minutes to travel the 150 million kilometres to Earth from the Sun's surface, yet the energy it brings with it, which started out only 700000 kilometres further away, in the Sun's centre, was released a million years previously. On its way to the solar surface, this energy has been absorbed and reemitted in a myriad interactions in the Sun's hot, gaseous interior. The light that reaches us is only remotely connected with the central processes that keep the Sun burning.

Were this the whole story, the centre of the Sun would be forever hidden from view, but fortunately it is not. A small fraction of the Sun's energy, about 2 per cent, exists in a different form, which can travel through the gassy mass as if it were not there at all, taking only three seconds to reach the surface from the centre. This energy is in the form of neutrinos. Largely unaffected by the Earth, the solar neutrinos stream through us constantly, raining down on us by day and up through the Earth by night. Most exciting of all, they bear the imprint of their birth at the heart of the Sun. They provide our only mirror on the Sun's core, if only we could see into it!

The Sun's energy comes from the conversion of hydrogen to helium. The basic reaction is between two single protons (hydrogen nuclei) to form a deuteron; the deuteron can then react with another proton to form helium-3. It then requires two helium-3 nuclei to react to make the stable form of helium, helium-4, which has a nucleus of two neutrons and two protons. Note that the first two steps in the chain must occur twice for every time the third step happens.

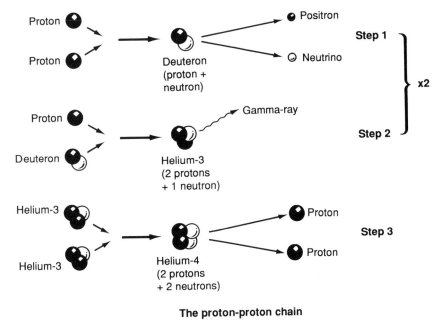

The proton-proton chain

The solar furnace

> A summer bather at a Cape Cod beach is essentially enjoying the power which comes from a titanic nuclear reaction eight light-minutes away.[3]
>
> *Hans Bethe, 1980.*

The Sun burns by converting the nuclei of the lightest element, hydrogen, into those of the second lightest element, helium. Each time this process happens, it releases an energy of 25 MeV, barely enough to move an amoeba, but sufficient to keep the Sun shining when multiplied up by the huge number of reactions that occur each second.

The origin of the energy lies in the mass of the original hydrogen nuclei. The conversion of hydrogen to helium in the Sun passes through several steps, but the net effect is to use four hydrogen nuclei (single protons) to make a nucleus of helium, consisting of two protons and two neutrons. In the helium nucleus, the four constituents are bound tightly together, so much so that the mass of the helium nucleus is less than the total mass of the four original protons. The difference is the 25 MeV, which is carried away mainly as kinetic energy by additional particles that participate in the overall reaction.

One of the most remarkable features of this hydrogen burning is that it depends on the most feeble of nuclear processes, the weak interaction. The first step in the chain proceeds through a process akin to beta-decay. A proton transmutes into a neutron near enough to another proton for the resulting neutron and proton to bind together to make a deuteron, the nucleus of the heavy form of hydrogen known as deuterium. In the transition from proton to neutron, a positron emerges, carrying the positive charge that originated with the proton, while a fourth player slips invisibly away. An electron-neutrino, with an average energy of 0.26 MeV, sets off on its lonely journey from the centre of the Sun.

However, even before the weak force comes into play, the initial two protons must come close enough together and to do this they must somehow overcome their natural electrical repulsion. This happens only through a quantum process known as 'tunnelling'; the thermal energies of the protons, even at the temperatures of 10 million degrees at the Sun's centre, are not sufficient to overcome the electrical repulsion through 'brute force' alone.

That the Sun burns at all therefore depends on two factors: quantum tunnelling and the weak interaction. That it is burning now is due to the rarity of both these processes. The probability for the basic fusion reaction is very, very small. Although the Sun's core is 10 times as dense as lead, a proton can exist there on average for 10 billion years before it joins with another to form a deuteron. However, we should be thankful that the process is so slow. Were the Sun to burn a little faster it might be already in its death throes, its hydrogen completely burned since the Solar System formed around 4.5 billion years ago!

Even though the Sun is burning relatively slowly, billions upon billions of electron-neutrinos should swarm out from its core each second. Indeed, knowing the Sun's measured luminosity (3.86×10^{26} joules per second) and that around 25 MeV (4×10^{-12} joules) is released in forming each helium nucleus, you can calculate that roughly 10^{38} conversions occur each second. Each conversion must release *two* neutrinos, as two protons must change to neutrons in forming a helium nucleus, so the total number of neutrinos is around 2×10^{38} per second. A tiny fraction of these – about 35 in every 10^{11} – should reach the Earth, and pass through more or less unimpeded. This implies that each second some 7×10^{28} neutrinos fly in through the surface of the Earth facing the Sun, and out through the opposite surface; that is, around 6×10^{10} neutrinos per second through each square centimetre!

Can we detect these neutrinos? Can we use them to 'see' directly to the Sun's core? Yes – in principle. But there is a problem. The very reason that the neutrinos can arrive at Earth as messengers straight from the centre of the Sun makes them almost impossible to stop. The weakness of their interactions with matter allows them to escape rapidly from the Sun, but also provides an unrelieved headache for experimental physicists. You have to be particularly stubborn to dedicate your life to detecting neutrinos from the Sun.

One such person is Raymond Davis Jr. In 1967, he began a famous experiment that has since monitored the arrival of neutrinos from the Sun almost continuously. Until 1988, this was the only experiment to claim to have detected solar neutrinos, but in doing so it provoked a great deal of discussion and speculation. For in its 25 years existence, the experiment has

The detection of neutrinos with chlorine is a two-step procedure. In the first step, a nucleus of chlorine-37 absorbs a neutrino and then emits an electron as it turns into argon-37. The next step occurs later (with a half-life of 35 days): the argon nucleus captures a nearby atomic electron and reverts to the more stable chlorine-37. But this leaves a vacancy in the orbitals of electrons in the atom, and in moving to fill the hole another electron releases energy, which, in turn, knocks a third electron from the atom. Detecting these emitted electrons, which have a characteristic energy, provides evidence for the decay and therefore for the prior existence of argon-37.

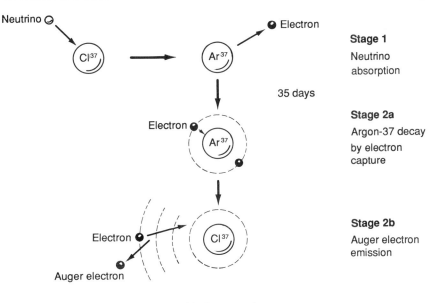

Stage 1

Neutrino absorption

35 days

Stage 2a

Argon-37 decay by electron capture

Stage 2b

Auger electron emission

The chlorine experiment

consistently measured the number of solar neutrinos to be less than the standard theory of the Sun predicts.

Ups and downs

> There was not much enthusiasm [in the early 1960s] among astronomers for what was viewed as an expensive experiment and not too much reason to hope that an observation could be performed that would actually detect solar neutrinos.[4]
> *John Bahcall and Raymond Davis Jr, 1982.*

Davis published his first findings about solar neutrinos in 1955. He was in the course of developing a detector to test whether neutrinos and antineutrinos are indeed distinct particles, in accordance with Dirac's theory for particles with spin $\frac{1}{2}$. Chapter 3 recounts how in the early 1930s Ettore Majorana had first suggested that it was possible that the neutrino and antineutrino could be the same particle, and how in 1949 Bruno Pontecorvo suggested a means for detecting the antineutrinos from a nuclear reactor in a large tank of carbon tetrachloride – dry-cleaning fluid.

Pontecorvo's idea was to look for a type of inverse beta-decay in which a neutron captures a neutrino and becomes a proton, while emitting an electron; if Majorana's hypothesis was correct, this reaction would also occur with antineutrinos. Pontecorvo found that chlorine-37 (with 17 protons and 20 neutrons), which forms nearly 25 per cent of all natural chlorine, would be a good nucleus to reveal this type of reaction.

If a neutron in the chlorine absorbed a neutrino, then it would change into a proton, and so change the nucleus into one of argon-37, which is chemically quite distinct from chlorine. Indeed, argon is an inert gas and so, Pontecorvo reasoned, it would become disconnected from the original carbon tetrachloride molecule, and could even be removed from the liquid. Moreover, the precise type of argon formed, argon-37, is radioactive. It has a half-life of 35 days before it changes back to chlorine-37 by capturing an atomic electron. The tell-tale sign of such a capture is the emission of a low-energy electron, evicted as the atom adjusts to the loss of one of its innermost electrons.

(What happens is that as another electron moves to fill the newly-created vacancy it releases energy which knocks out a third electron. In 1922, in her attempts to understand the lines in the spectrum from beta-decay (see Chapter 2), Lise Meitner correctly suggested that this effect gives rise to low-energy electrons produced as an atom reorganises itself after beta-decay. However, the effect is now named after Pierre-Victor Auger, the French physicist who discovered it again in 1925.)

Davis's first large chlorine detector contained 3800 litres of carbon tetrachloride (CCl_4) and was exposed to the antineutrinos from the research reactor at the Brookhaven National Laboratory on Long Island, New York. As part of his tests to understand the background due to cosmic rays, he buried the tank beneath 5.7 metres of the soil at Brookhaven, so as to reduce greatly the effects of nuclear particles in the cosmic-rays. The amount of argon-37 produced, if any, was smaller than he could detect. But from this result, Davis could calculate that the number of *detectable* neutrinos arriving from the Sun each second must be less than 10^{14} per square centimetre.

The key word here is 'detectable'. The neutrinos that emerge from the initial proton–proton fusion reaction have a maximum energy of only 0.42 MeV. However, the capture process in chlorine-37 can occur only if the neutrinos have an energy greater than 0.86 MeV. At the time – 1955 – Davis thought that his only chance was to detect the higher energy neutrinos that would be produced if the Sun converted its hydrogen to helium in a different way, in a chain of reactions that involves carbon, nitrogen and oxygen. But astrophysicists already believed that the Sun's interior was too cool for this chain to occur, and that the burning of hydrogen must take place through the proton–proton chain. If this were indeed so, then observing solar neutrinos appeared hopeless.

Over the next few years a succession of events occurred that led Davis's expectation for a solar neutrino detector to rise and fall, but eventually to culminate in his successful experiment in the Homestake Gold Mine in South Dakota. The first key was the discovery that a rare variation in the way that helium forms from four protons might occur much more frequently than anyone had previously thought.

In the second step of the reaction chain that leads to helium, the deuteron (proton plus neutron) formed in the first step binds with another proton to produce helium-3. (This is a light form of helium that contains one neutron less than the usual form, helium-4). Most of the time, the next step is for two nuclei of helium-3 to combine to form helium-4, emitting the two excess

One alternative way for helium-4 to form from four protons involves several steps via more complicated nuclei. About 0.1 per cent of the Sun's helium-4 is produced via the formation of boron-8, from helium-3 and helium-4 produced from the basic proton–proton reaction illustrated on p.149. The boron-8 is unstable and soon decays to beryllium-8, emitting a positron and a neutrino in each decay. These neutrinos have a much higher energy (up to 14 MeV) than those released in the initial proton–proton reaction (up to 0.42 MeV).

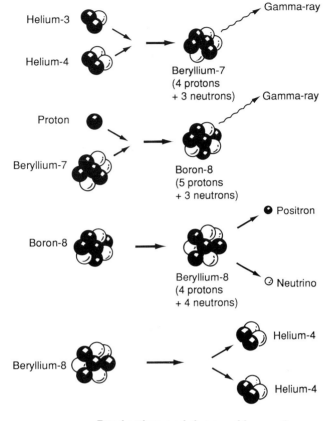

Production and decay of boron-8

protons, which carry away a substantial amount of the energy released overall. However, another possibility is for a nucleus of helium-3 to combine with one of helium-4, already made, to form beryllium-7.

In 1958, two physicists at the Naval Research Laboratory in Washington discovered that this reaction occurs 1000 times more frequently than anyone had previously suspected. Indeed, we now know that it occurs in 15 per cent of the conversions of hydrogen to helium. Willy Fowler, a leading nuclear astrophysicist at the Kellogg Radiation Laboratory at the California Institute of Technology, immediately realised the ramifications for detecting solar neutrinos.

Most of the time, any beryllium-7 produced in the Sun goes on to absorb an electron and so become lithium-7; a nucleus of lithium-7 can then capture an additional proton to form two helium nuclei, so providing the last step in the conversion of hydrogen to helium. But another possibility is for the beryllium-7 to absorb a proton, so forming boron-8. Boron-8 is unstable, and soon decays to beryllium-8 by emitting a positron, and a neutrino. What is so important, as Fowler realised, is that this neutrino's energy can be as high as nearly 15 MeV – well above the energy needed for absorption by chlorine-37.

Fowler, and also Alastair Cameron at Chalk River in Ontario, were quick to point out the possibilities to Davis. He, in turn, calculated that the two 1900-litre tanks of perchloroethylene (C_2Cl_4), which he currently had installed at the Savannah River Reactor in Georgia, would capture more than seven neutrinos a day!

However, the excitement that this discovery aroused was short-lived. By 1960, Ralph Kavanagh, at the Kellogg Laboratory, had measured the probability for a beryllium-7 nucleus to absorb a proton, and his result was disappointingly low. Fowler estimated that only about 0.02 per cent of the conversion chains in the Sun would go via this reaction, which implied that the number of higher energy neutrinos reaching the Earth would be about 10 million per square centimetre per second – too low a rate for a viable experiment. Reviewing the situation in 1960, Fred Reines wrote:

. . . a flux of around $10^7/cm^2$sec is so ineffective that a detection volume in excess of 1000 cu.m would be required for 10 counts/day. Detectors of this size are considered impractical . . .

. . . the probability of a negative result even with detectors of thousands or possibly hundreds of thousands of gallons of CCl_4 tends to dissuade experimentalists from making the attempt.[5]

Davis, however, was not so easily deterred, 'even though the prospects for observing solar neutrinos looked dim'.[6] He believed that it would not be unfeasible to scale up by a factor of a hundred the 3800-litre experiment he had used at Savannah River. Then in 1963 the tables turned again. This time the good news concerned reactions not in the Sun, but in chlorine-37!

By 1962, Davis had begun a collaboration with an astrophysicist at the University of Indiana, John Bahcall, which was to last until this day. In the summer of 1962, Bahcall joined Fowler's group at the Kellogg Laboratory to work on a detailed model of the Sun with a view to calculating the number of neutrinos reaching the Earth. A year later, he made the second discovery that helped to make a solar neutrino detector a realistic proposition. He found that chlorine-37 nuclei should capture the neutrinos from boron-8 twenty times more readily than had been previously supposed. The chlorine-37 could make transitions not only to the normal, ground state of argon-37, but also to excited states in the argon nucleus. This raised Davis's hopes tremendously:

The realization that neutrino capture to the analog state in ^{37}Ar greatly increased the total capture rate made an enormous difference in Davis's view of a 100,000 gal experiment. It seemed to him that the analog state was a beautiful new concept in the present context that *should* appeal to nuclear physicists. Moreover, the total expected capture rate was increased to about 4 to 9 per day, making the experiment seem more reasonable . . . [and] the chlorine experiment could be considered as a way of measuring the central temperature of the Sun, which Davis felt should appeal to astrophysicists.[7]

Raymond Davis Jr (right) and John Bahcall (centre) photographed shortly after their proposal in 1964 that detecting solar neutrinos with chlorine-37 would be feasible. They are standing together with Don Harmer, who also played a leading part in the experiment, in front of a small version of the chlorine tank. (Courtesy R. Davis Jr and J. N. Bahcall.)

The rate of the reactions producing the boron-8, and hence the number of energetic neutrinos, depended very much on the temperature of the solar interior. Indeed, Bahcall claimed that a measurement of the neutrino flux accurate to 50 per cent would reflect the temperature of the Sun's core to within 10 per cent.

In March 1964, Bahcall and Davis published papers in *Physical Review Letters* on the theory and practice of a proposed 380 000-litre experiment. Later in the year, Willy Fowler put his weight behind the proposal, and in December the Homestake Company provided a 'favourable' estimate of the costs of excavating a suitable cavern nearly 1500 metres below ground at their gold mine in Lead, South Dakota. The year culminated with the discovery of calcium-37, with a half-life close to that which Bahcall had predicted in the same set of calculations that revised the capture rate in chlorine-37. He later referred to receiving this news as 'the most exciting and satisfying single moment of his professional career'.[8]

Down the Homestake Gold Mine

Ray Davis tells me that the experiment is simple ("Only plumbing") and that the chemistry is "standard". I suppose I must believe him, but as a non-chemist I am awed by the magnitude of his task and the accuracy with which he can accomplish it.[9]

John Bahcall, 1969.

In May 1965, the Homestake Mining Company began to excavate the cavern, $9 \times 18 \times 9.6$ cubic metres, which would house the 380 000-litre tank of C_2Cl_4. About two months later they were finished, and Davis and his colleague Blair Munhofen, who had done much to help find a suitable mine, had their first viewing:

They were brought into the room and immediately started looking around with miner's lamps. Suddenly, the lights were turned on and they could see the enormous room with its walls covered with chain-link fencing, the concrete floor with pedestals for the tank supports and the monorail for the lifting hoist 32 ft above.[10]

The tank itself was built by the Chicago Bridge and Iron Company; the company was evidently 'intrigued by the aims of the project and the unusual location'.[11] There were two important features in the construction. One was that it had to be built from steel plate that did not have too high a natural emission of alpha-particles from its surface. Such alphas could have produced argon-37 in the detector, so Davis, Munhofen and Don Harmer checked the plate themselves before giving the go-ahead for the construction. The other important feature was that the tank should be extremely leak-tight, so as to avoid any argon leaking in from the atmosphere. In this respect, the company had great expertise in building large leakproof vessels for the space agency, NASA.

The tank was complete by the summer of 1966, and eventually the filling began, with C_2Cl_4 from the Frontier Chemical Company in Wichita, Kansas, provided in 2500-litre tank wagons. Like the steel plate, each tank-load of liquid was sampled and tested for alpha-emission before it left Wichita. The filling took five weeks. Finally, once the system for processing the C_2Cl_4 had been installed, Davis's team began the long task of removing dissolved air from the 380 000 litres of liquid, so as to minimise the amount of atmospheric argon. Then the detector was finally ready to lie in wait for neutrinos from the Sun.

And so began a procedure that has continued, with few breaks, for more than 20 years. The technique is to allow any argon-37 to accumulate in the tank over a period of one to three months. Then the tank is flushed with about 10 000 litres of helium gas per minute, to remove as much of the argon from the liquid as possible – about 95 per cent. After passing through the tank, the helium flows eventually through a charcoal trap which is cooled to 77 K by liquid nitrogen. At this low temperature, the argon gas condenses out into the trap, leaving the helium gas, which condenses at a far lower temperature, to be recycled through the tank.

Once the flushing is finished, the charcoal trap is heated to drive off the argon gas, which is collected and purified chemically to remove any traces of other unwanted radioactive elements. The final sample of gas, about half a cubic centimetre, consists mainly of argon-36, a stable form of argon that the team introduces deliberately at the start of each period of data collection. By measuring how much argon-36 is removed, the researchers can check the efficiency of the extraction system each time. In addition, there may be a small amount of argon-40, the common form of argon that occurs in the atmosphere. Finally, the sample should contain some tens of atoms of argon-37, produced by the interaction with chlorine-37 of neutrinos from the decays of boron-8 at the heart of the Sun.

Raymond Davis and his colleagues eventually built a chlorine-37 solar neutrino detector 1500 metres below ground in the Homestake Gold Mine at Lead, South Dakota. The main body of the detector is a tank 14.5 metres long and 61 metres in diameter, which holds 380 000 litres of perchloroethylene. It is seen here with Davis on the catwalk and technician John Galvin below. The volume surrounding the tank can be filled with water to provide a shield against neutrons during running. (Brookhaven National Laboratory.)

The next step is to count the number of argon-37 atoms in the purified argon sample. To do this, the team must place the entire sample in a small proportional counter. This contains a tiny metal tube, about 2.5 centimetres long and 0.4 centimetres in diameter, with a wire running along the centre. A high voltage applied between the wire and the tube sets up an electric field in the gas within the tube. Any charged particle passing through the gas ionises it, releasing electrons which move in the electric field towards the central

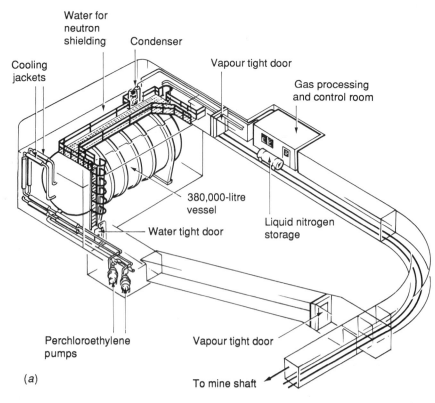

Cooling jackets

Water for neutron shielding

Condenser

Vapour tight door

Gas processing and control room

380,000-litre vessel

Water tight door

Liquid nitrogen storage

Perchloroethylene pumps

Vapour tight door

(a)

To mine shaft

(a) Every few weeks, the per-chloroethylene in the tank in the Homestake Mine is flushed through with helium gas to remove any argon. The helium is drawn off from the top of the tank and the argon filtered off and purified in the gas-processing room. The system for circulating helium gas through the liquid, to extract the argon-37, is a vital element of the detector.
(b) Gerhart Friedlander inspects the pumps, as Davis enters by the door.

(b)

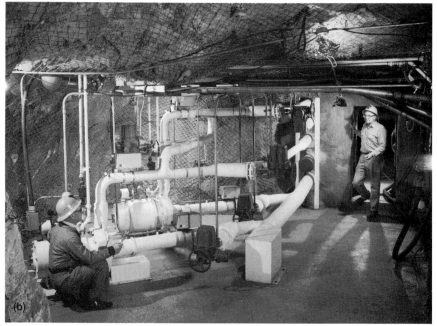

The observations made by Davis and his colleagues over a period of nearly 20 years, consistently show that one atom of argon-37 is produced in the detector in the Homestake Mine about every two days. The standard solar model, however, predicts that argon-37 should be made at three times this rate, as indicated by the horizontal line.

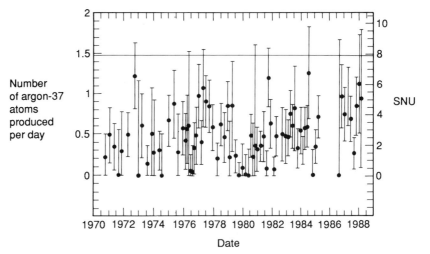

wire – the positive electrode. These electrons can also ionise the gas, causing an avalanche effect, so that eventually a large number of electrons arrives at the wire. The total number of electrons collected at the wire, and therefore the size of the electrical pulse produced, is proportional to the energy of the original charged particle – hence the name, 'proportional' counter.

When argon-37 decays, it releases low-energy, Auger electrons, which travel for only about 0.1 millimetres through the gas in the proportional counter. This means that all the ionisation electrons are produced close together, and all arrive at the central wire within a short space of time, giving a sharp edge to the detected pulse. This characteristic pulse shape helps the team to distinguish pulses due to the decays of argon-37 from those due to other sources. For a typical sample, the team counts about six pulses of this kind. In other words, they can detect a few individual atoms of argon-37 culled from a tank containing in the region of 10^{30} atoms – a remarkable achievement by any standard.

The experiment has now run almost continuously for more than 20 years. In May 1985, operation stopped for about a year after both pumps that circulate the liquid had failed. One new pump was installed in 1986, and in October of that year the experiment began again. In April 1988, a second new pump was added to the system.

The results have provided particle physicists and astrophysicists with one of the biggest puzzles of the second half of the twentieth century. Right from the start, when the first results were published in *Physical Review Letters* in 1968, Davis's detector has consistently found fewer solar neutrinos than Bahcall and his colleagues predict from their detailed models of the solar interior. During some 80 periods of data collection, from 1970 to 1988, the detector captured an average of one neutrino every two to three days. This is only a quarter or so of the captures expected from the theoretical predictions.

As Bahcall and Davis commented in 1982,

It is surprising to us, and perhaps more than a little disappointing, to realize that there has been very little quantitative change in either the observations or the standard theory since these [first] papers appeared, despite a dozen years of reexamination and continuous effort to improve details.[12]

What can be wrong? Is it the experiment? Is it the theory of the Sun? Or is it our understanding of neutrinos? Davis and his colleagues have taken inordinate care to understand their experiment, and it has passed the sceptical scrutiny of fellow experimenters world-wide, so that few, if any, dispute the results. Instead, the results have set a challenge that a number of physicists have been unable to resist, leading them to invent intriguing new theories and ingenious new experiments.

The second solar neutrino experiment

> Just a few months ago, results from a directionally sensitive experiment gave clear-cut evidence that the Sun is emitting neutrinos – the first experimental evidence that the Sun's energy indeed originates in nuclear reactions.[13]
>
> *Lincoln Wolfenstein and Eugene W. Beier, 1989.*

Between 1970 and 1988, the rate of neutrino captures per atom of chlorine-37 in the tank in the Homestake mine came to $(2.2 \pm 0.3) \times 10^{-36}$ per second, where the various uncertainties in the measurement have been brought together to yield the overall 'error' of 0.3×10^{-36}. Because the captures are so rare, the units involved are rather cumbersome, so in 1969 John Bahcall proposed a more convenient 'solar neutrino unit' equal to 10^{-36} captures per second per atom of chlorine-37. The name of the unit is usually abbreviated to the handy anagram SNU, pronounced 'snew'.

The problems arise when the experiment is compared with predictions based on the 'standard solar model' in which hydrogen fusion fuels a steadily evolving Sun. This model predicts a capture rate in the chlorine experiment of 7.9 ± 2.6 SNU, which in no way overlaps with the measured rate, even if the uncertainties are taken to their extremes!

Is there a problem with the experiment? After some 20 years of scrutiny, the answer must almost certainly be 'no'. However, scientists often doubt the results of one experiment until another has independently corroborated the findings, and for 20 years, the experiment in the Homestake Gold Mine remained unique. No one else seemed willing to invest the necessary time and money. Then at last, in 1989, support emerged for Davis's results from an entirely different kind of experiment, with the added ability to show that it does indeed detect neutrinos that originate with the Sun!

This experiment sits in the Kamioka metal mine, half way round the world from Dakota in the Japanese Alps, 300 kilometres west of Tokyo. The detector here was originally set up by a team from several Japanese Universities and the National Laboratory for High-Energy Physics (KEK) at Tsukuba. The initial aim was to search for the possible decay of the proton, hence the name Kamiokande, for Kamioka *n*ucleon *d*ecay *e*xperiment. Then in 1984, the Japanese invited a group from the University of

The Kamiokande II detector, 1000 metres underground in the Kamioka Mine in the Japanese Alps. The main inner detector is a tank of water 16 metres high and 15.6 metres diameter. Its inside surfaces are covered with a regular array of light-sensitive phototubes, each 50 centimetres diameter. This view is from the top edge of the inner detector, with a person just visible at the bottom. (M. Koshiba, Kamiokande II.)

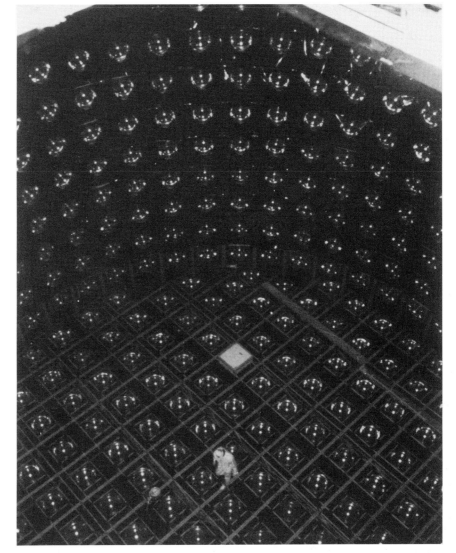

Pennsylvania to join the original team, to modify and improve the detector so as to make it sensitive to solar neutrinos – and Kamiokande II was born.

Kamiokande uses a cheap and simple medium to detect solar neutrinos: water. In a game of subatomic billiards, in which the onlookers need almost infinite patience, neutrinos from the Sun can occasionally collide with an electron in the water, knocking it forwards, close to the neutrino's original direction. As the electron moves through the water it creates an electromagnetic 'wake', producing a cone of light about its path. Light-sensitive phototubes arrayed around the surfaces of the detector intercept this cone of light and so reveal the electron. The axis of the cone gives the electron's direction; the intensity of the light gives a measure of the electron's energy; and the time at which the light signals are detected can be recorded. So a water detector has the capacity to measure as much as possible regarding the interactions of neutrinos. However, like the chlorine experiment, Kamiokande II can reveal only the higher-energy solar neutrinos, because only electrons with relatively high energies (greater than 9.3 MeV in the first results) produce useful signals.

A major difficulty with a water detector is that single electrons can arise in many other ways, especially from the beta-decays of radioactive nuclei and from the interactions of gamma-rays and neutrons due to radioactive materials. However, by the beginning of 1986, after a year of dealing with a number of sources of background reactions, Kamiokande II was ready to commence in earnest the search for signals from solar neutrinos. The expected rate was of one solar neutrino interaction every few days, although the phototubes triggered every couple of seconds, about half the time because they were picking up light produced by cosmic-ray muons, even deep underground! However, on analysing the recorded patterns of light, the physicists could reject the signals probably due to anything other than low-energy neutrinos.

In July 1989, the team published a paper containing two long-awaited results. First, they had found signals that appeared to be due to solar neutrinos, in which the single electron detected pointed back towards the Sun more often than it pointed in other directions. Secondly, it transpired that Kamiokande II also finds fewer higher-energy neutrinos than the standard solar model predicts.

By April 1990, the Kamiokande team had results from more than 1000 days of data taking. These showed that the detector had observed 0.46 times the number of neutrino interactions predicted on the basis of decays of

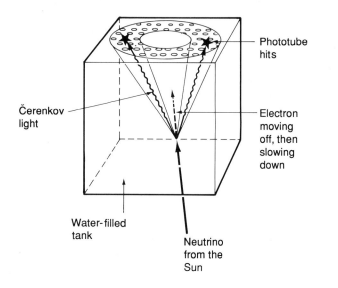

A neutrino from the Sun can interact with an electron in water, knocking it forwards in almost the same direction as the neutrino was travelling. As the electron moves through the water it radiates a cone of light about its path – Čerenkov radiation – which fires a ring of phototubes when it reaches the walls of a detector.

Results from the chlorine detector in the Homestake Mine suggest that the number of solar neutrinos detected may decrease, the larger the number of sunspots there are on the surface of the Sun.

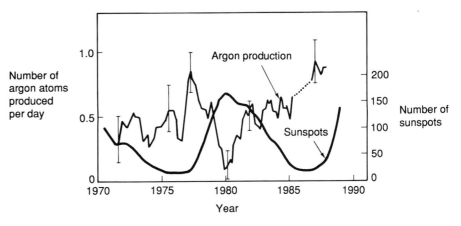

boron-8 in the standard solar model. Moreover, the team found that the energy spectrum of the scattered electrons closely matched what would be expected if boron-8 decay were indeed the source of the neutrinos. In a paper published in *Physical Review Letters* in September 1990, they concluded that:

. . . the mechanism of energy generation in the Sun based on the fusion reactions which give rise to ^{8}B as a by-product would appear to be unequivocally confirmed by the detection of neutrinos which could only have originated in the core of the Sun.[14]

It is not a straightforward matter to compare the ratio between experiment and theory measured in Kamiokande II with a similar figure for the chlorine experiment. Chlorine can capture any neutrinos with an energy greater than 0.86 MeV, so it is necessary to make some assumptions about what proportion of the neutrinos detected in chlorine come from boron-8. However, it does seem that Kamiokande II detects rather more neutrinos than does the chlorine experiment. John Bahcall and Hans Bethe have argued that the number of neutrinos detected in Kamiokande would imply a rate of 4 SNU in a chlorine detector, which is well outside the number measured in the Homestake Mine of 2.1 ± 0.3 SNU. This difference could be a pointer to exciting new theories, not about the Sun, but about the neutrinos themselves, as we shall see later in this chapter.

Meanwhile, the Kamiokande detector should continue to provide a useful comparison for at least a few years to come. It should also help to cast light on an apparent twist in the solar neutrino mystery that has recently emerged. Since late 1985, the experiment in the Homestake Mine has tended to detect more solar neutrinos than in previous years, with the capture rate rising to 3.6 ± 0.7 SNU. Is the Sun responsible for this apparent change? Plotted over the years, the data from the chlorine detector have fluctuated, and it is possible that there may be some connection with the number of sunspots. These dark, cooler regions on the Sun's visible surface fluctuate regularly in number over a period of 11 years – the so-called solar cycle. The results from the Homestake Mine seem to show that more neutrinos are detected during the times when there are fewer spots on the Sun, and in 1986 the sunspot cycle was at a minimum. However, the 1000 days of data from Kamiokande II, collected over a time when the sunspot cycle was changing rapidly from a minimum to a maximum, show no variation at all.

Not everyone is convinced that there is a definite correlation between sunspots and solar neutrinos, but time should provide an answer, if the experiment continues to run through the sunspot maximum around 1991. Will the capture rate in the chlorine experiment fall again, as the number of sunspots increases? Bahcall (who is sceptical) and Davis have wagered a bottle of champagne on the effect. Perhaps it will have been won by the time you read these words!

Trouble with the Sun

> The surprising discrepancy between the calculated and the observed rate of solar neutrino captures in chlorine (SNUs) is a long-standing puzzle that has spawned many imaginative scientific explanations and several science-fiction novels.[15]
> *John Bahcall, 1987.*

Is there a problem with the Sun, or more precisely with our understanding of it? John Bahcall, at the Institute for Advanced Study in Princeton, New Jersey, since 1968, and Roger Ulrich from the University of California at Los Angeles, have led the efforts to produce a standard solar model. In their calculations, they begin by making the basic assumption that the Sun is in hydrostatic equilibrium, with the outward pressure, due to the energy produced in nuclear reactions, balancing the inward pressure due to gravity. Their technique is to write down the various equations that describes the Sun's evolution and then to solve them for different times, from the early days of the Sun through to its present age of 4.5 billion years.

The aim is to vary slightly the starting conditions, and perhaps some of the parameters that govern the Sun's evolution, and then discover exactly which variation leads to a Sun with the correct present size, age and luminosity. This particular variation then yields the standard solar model. Among other things, Bahcall and Ulrich allow the initial proportions of hydrogen, helium and other elements to vary in their various models. They find that the version that gives the correct properties for the Sun at its present age is the one that starts with a composition similar to that observed for the Universe in general. This implies that their technique works, and provides a model of the Sun that reflects reality. Why then does it predict too high a capture rate for solar neutrinos in the chlorine detector?

One of the major problems with the chlorine detector is that it is sensitive to only a small percentage of the neutrinos that stream from the Sun – those with the highest energies, which are produced mainly in the decays of boron-8. Most attempts to resolve the discrepancy between the measured and predicted capture rates seek to the lower the number of boron-8 decays and so reduce the number of high-energy neutrinos.

The solar furnace makes boron-8 when nuclei of boron-7 capture an additional proton. But this process depends strongly on the temperature of the furnace. There is a large electrostatic repulsion between a proton and a boron-7 nucleus with seven positive charges. The temperature of the Sun's interior is far too low for thermal energy to help the protons overcome this repulsion, so the only way the capture can proceed is via quantum mechanical 'tunnelling'.

According to quantum theory, the proton's energy fluctuates about its average, thermal value, and this means that it has a very small but finite chance of occasionally having enough energy to cross – tunnel through – the repulsive barrier around the boron-7 nucleus. This small chance depends very much on the temperature, and a relatively small reduction in the temperature inside the Sun could significantly reduce the rate at which boron-8 forms. And this, in turn, would reduce the number of solar neutrinos captured in the Homestake Mine.

For this reason, attempts to adjust the solar model have often centred on reducing the temperature of the interior. For example, can heat escape from the centre more rapidly than is usually assumed? This would be the case if the Sun's interior had smaller amounts of the heavier elements than at the surface. Heavier elements absorb more energy in collisions with photons, so if there were a smaller proportion of them nearer the centre of the Sun, more energy would escape to the outer layers. The result would be a cooler centre.

Another proposal to reduce the temperature at the centre hinges on the existence of 'WIMPs', or weakly interacting massive particles. Astrophysicists have proposed the existence of such particles to solve a different problem, that of the Universe's 'dark matter'. Observations of the motion of stars in galaxies suggest that as much as 90 per cent of the mass of a galaxy is hidden from view because it does not radiate. There have been many speculations for the source of this mass, including a halo of massive neutrinos, as we shall see later. The existence of some new kind of massive particle that interacts only very weakly with matter is but one possibility.

WIMPs with the right properties – for example, a mass between 2 and 10 GeV – would have the attraction of not only providing the galactic missing mass, but of also solving the solar neutrino problem by carrying away energy from the Sun's core, thereby reducing the temperature of the inner layers. But the very same results from CERN's electron–positron collider, LEP, which limit the possible number of neutrino types (see Chapter 4) also put severe constraints on the nature of WIMPs. It seems that the various WIMPs suggested so far to solve the solar neutrino problem are ruled out by the results from LEP.

There have been many other suggestions for nonstandard solar models. But none of them is entirely free from difficulties. While they may solve the solar neutrino problem, they often raise other problems, indicating that the solution to the mystery of the missing solar neutrinos lies elsewhere. That leaves one more possibility. Is there something wrong with the neutrinos?

Trouble with neutrinos

[The MSW effect] is such a beautiful phenomenon that Nature would be well advised to use it. After all, it may eventually give us the unambiguous, incontrovertible, uncontestable, clear and definitive evidence we so eagerly seek that the neutrino has mass.[16]

Peter Rosen, 1986.

Since the 1950s, when Davis first conceived of his solar neutrino experiment, particle physicists have found that there are three distinct types of neutrino: electron-neutrinos, muons-neutrinos and tau-neutrinos. But the nuclear interactions in the Sun emit only one of these types, electron-neutrinos; likewise, the chlorine-37 experiment can detect only electron-neutrinos. The possibility that neutrinos might be able to change from one type to another, as discussed in Chapter 4, would therefore provide a neat explanation for the missing solar neutrinos.

The basic idea behind such neutrino 'oscillations' is that the neutrino states we observe are quantum mixtures of underlying 'base' states. And if these base states differ slightly in mass, then the mixture fluctuates as it propagates through space. The effect at the quantum level is analogous to what happens with two 'coupled' pendulums, swinging from the same horizontal string. If you set one pendulum in motion, it will gradually transfer its motion to the other and come to a temporary halt before picking up its swing again at the expense of the second pendulum.

Suppose, then, that as the electron-neutrinos produced in the Sun travel the 150 million kilometres to Earth, some of them change to muon-neutrinos or to tau-neutrinos. Any of these altered neutrinos that arrive in the Homestake Gold Mine will have no chance at all of being captured by the chlorine-37 in the tank there, and will instead remain completely hidden from view.

It is indeed possible to explain the shortfall in the detected solar neutrinos in this way, but it would be very surprising if mixing in its simplest form were the correct explanation. Quantum mixing occurs in other areas of particle physics, notably in the decays of certain quarks through the weak force. But to reproduce the observed reduction in solar neutrinos requires that there is an unprecedented degree of mixing between the neutrino base states as they travel through free space – far greater, for example, than the mixing required to understand the weak interactions of quarks.

However, this is not the end of the story. In 1978, Lincoln Wolfenstein, at the Carnegie Mellon University in Pittsburg, realised that neutrino oscillations should be modified by matter. This is because at the low energies of the solar neutrinos, electron-neutrinos have a means of interacting with electrons in ordinary matter that is not open to muon-neutrinos or tau-neutrinos.

The only way that neutrinos can interact with other particles is through the weak force, or in other words through the exchange of W^+ and Z^0 particles (see Chapter 5). All types of neutrino can interact equally with the

Any type of neutrino can interact with electrons or protons in the Sun by exchanging a Z^0 particle – that is, through the weak neutral current. But the only interaction possible via the weak charged current is between electron-neutrinos and electrons. Solar neutrinos do not have enough energy to create the muon or the tau that would have to be produced by the charged current, and only antineutrinos, not neutrinos, should interact with protons in this way.

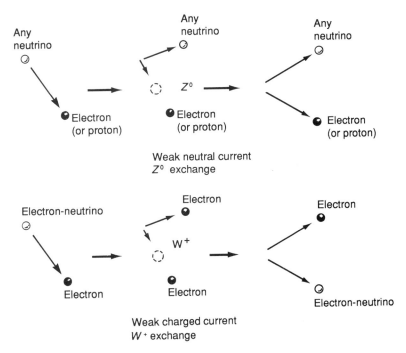

electrons and nucleons in matter through the exchange of Z^0s, in 'neutral-current' reactions. However, another class of reaction comprises the 'charged-current' interactions, which involve the exchange of a charged W particle. In this case, charge changes hands, so the neutrino changes into the related charged particle – an electron for an electron-neutrino and so on. However, at energies typical of solar neutrinos, only electron-neutrinos can react with electrons in this way. The related reactions involving muon-neutrinos or tau-neutrinos do not occur, because the neutrinos do not have enough energy to produce the more massive charged muons or taus.

The net effect of this additional interaction for electron-neutrinos is that any mixing depends on the density of electrons in the matter through which the neutrinos are travelling. The centre of the Sun is very dense – ten times as dense as lead – whereas at the surface the density drops close to that of space. In 1985, S.P. Mikhayev and Alexei Smirnov at the Institute for Nuclear Research in Moscow, realised that as neutrinos travel from the centre the degree of mixing will change, and should in fact pass through a maximum along the route to the surface. And this could be enough to change the original electron-neutrinos into muon-neutrinos.

In this way, Smirnov and Mikhayev were able to explain neutrino oscillations without the need for an unnaturally large amount of mixing. The intrinsic mixing would be small; it would be the enhancing effect of the electron-neutrino interactions with electrons that would make the mixing sufficient to change neutrinos from one state to another.

The MSW (Mikhayev–Smirnov–Wolfenstein) effect captured the imagination of many physicists as being perhaps the most plausible reason for the missing solar neutrinos. Unlike some other explanations it does not demand completely new physics, such as the existence of WIMPs. Moreover, it meshes well with existing attempts to extend the Standard Model to a 'grand unified theory', or GUT. Such a theory would deal simultaneously with strong and electroweak interactions, rather than handling them in the separate compartments of quantum chromodynamics and electroweak theory. And the hope is that a viable GUT would cast light on questions that the Standard Model fails to address, such as why the masses of particles are what they are, and why there are three 'generations' in the two families of quarks and leptons.

In GUTs, quarks and leptons fall into the same mathematical symmetry group. One result is that all quarks and leptons, including the neutrinos, must have some mass. Another is that transitions between quarks and leptons are possible. Such transmutations could be manifest in the decay of a proton to a positron, say albeit on a very long time scale due to the weakness of the underlying force. The feebleness of these reactions would, in turn, be related to the large mass of the carrier particle involved, which could be around 10^{15} GeV. This would be the energy at which the distinction between strong and electroweak forces would fade away, rather as electromagnetic and weak effects take on equal roles at energies around 100 GeV, the mass of the W and Z particles.

In the Standard Model, neutrinos have no mass, and so no kind of neutrino oscillations can occur. If the MSW effect proves to be really at work on solar neutrinos, it will provide a valuable key to unlocking physics beyond the Standard Model. In John Bahcall's words:

... information about the grand unification mass scale, [around] 10^{15} GeV, [will be] being obtained from an interaction that is driven by a neutrino mass difference of order 10^{-20} GeV. How marvelous and awesome a possibility![17]

The gallium solution

> Of particular interest is the use of Ga^{71} as a target . . . An observed capture rate significantly below the predicted rate would be a strong indication of some kind of neutrino transformation.[18]
>
> *Lincoln Wolfenstein and Eugene W. Beier, 1989.*

As long ago as 1965, when the building of Davis's detector was just beginning, Vladimir Kuzmin at the P.N. Lebedev Institute in Moscow suggested an alternative technique for trapping solar neutrinos. Kuzmin's proposal was to use gallium-71, rather than chlorine-37, to catch the elusive neutrinos. The big advantage would be that gallium-71 could capture neutrinos with energies as low as 0.23 MeV. In other words, it could pick up neutrinos emitted in the basic proton–proton interactions that start the conversion of hydrogen to helium. The big disadvantage was that gallium is by no means a common material like chlorine. In 1965 the idea of using tens of tonnes of the stuff in a neutrino detector seemed high impractical, if not

Most of the neutrinos from the Sun will be low-energy neutrinos produced in the basic proton–proton interaction; but there should also be neutrinos produced at two specific energies when beryllium-7 captures an electron, and others from a relatively infrequent proton–electron–proton reaction, as well as the rare high-energy neutrinos from the decays of boron-8. The chlorine detector is sensitive to only the higher-enery neutrinos; the gallium detectors should register the low-energy neutrinos produced in the initial proton–proton reactions.

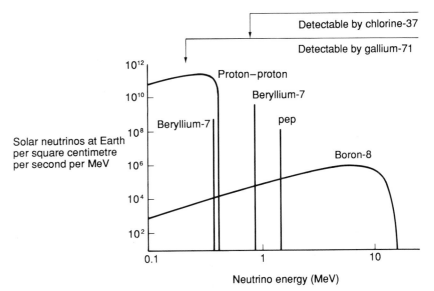

downright impossible. As Bahcall has pointed out, at the time 'the required amount of gallium exceeded the annual world production by an order of magnitude'.[19]

But in the microelectronic revolution of the 1970s, gallium asserted itself in a new role in the semiconductor industry, in compounds that could improve on the ubiquitous silicon in special applications. By the end of the decade the production of large quantities of gallium no longer seemed totally out of the question, and several researchers began to think seriously about the possibilities of a gallium solar neutrino detector.

Then, in 1990, 25 years after Kuzmin had made his proposal, the first results from a gallium detector were announced – and they took almost everyone by surprise. The new detector found far fewer neutrinos than the standard solar model predicted; indeed, it seemed almost oblivious to the neutrinos from the basic proton–proton reaction! The possibility that tantalised everyone was that they were seeing the first strong evidence for neutrino oscillations of the kind proposed by Mikhayev, Smirnov and Wolfenstein.

Appropriately enough, the first large-scale gallium detector to begin collecting data has been built in the Soviet Union. Not far from Mount Elbrus (5642 metres), the highest peak in the Caucasus mountains on the frontier between Europe and Asia, lies the Baksan Neutrino Observatory, in a cavern excavated under Mount Andyrchi along the Baksan Gorge. This is the home for SAGE, the Soviet–American Gallium Experiment. It has been built by a team from the Institute for Nuclear Research in Moscow, led by V.N. Gavrin and Georgii Zatsepin, together with American groups from the Los Alamos National Laboratory and the universities of Pennsylvania (including Ray Davis), Louisiana State and Princeton.

Gallium is a metal, unique in being a liquid over a wide range of temperatures, from 30°C to more than 2000°C. It occurs naturally in two forms: gallium-69 (31 protons and 38 neutrons) and gallium-71 (31 protons and 40 neutrons). It is the gallium-71, which forms 40 per cent of natural gallium, that is useful for catching neutrinos. When a gallium-71 nucleus captures an electron-neutrino, one of its neutrons converts into a proton to give a nucleus of germanium-71 (32 protons and 39 neutrons). However, the germanium produced in this way decays, with a half-life of 11 days, when one of the protons captures an electron and the germanium-71 reverts to gallium-71. Just as in the decay of argon-37, produced in the chlorine detector, the capture sets off the emission of an Auger electron as the new nucleus settles down to a stable state.

In the SAGE detector, the gallium is kept molten in reactor vessels, at around 30°C. At the start of an 'exposure', a known amount of nonradioactive germanium is added to the detecting liquid. Some 20–30 days later, the researchers remove this germanium, together with any germanium-71 that has been produced. To extract the germanium, they add a solution of hydrochloric acid and hydrogen peroxide in water to the molten metal in the

The Baksan Gorge, bounded to one side by the steep tree-covered slopes of Mount Andyrchi. Various underground laboratories are arranged along a tunnel more than 4 kilometres long, which begins in the 'wall' above the building near the centre of this picture. The SAGE solar neutrino detector is located 3.5 kilometres into the mountain, 2 kilometres beneath the summit. (D. Wark/SAGE.)

(a) A view across the underground chamber housing the reactor vessels for the SAGE experiment, shows clearly the stirring motors attached to their tops. There are 10 vessels altogether, but only four of those on the right have been used in the initial work with 30 tonnes of gallium. The intricate chemical extraction system, to the right of the vessels, is seen more clearly in (b). (T. Bowles/SAGE.)

reactors and 'stir well' for about 15 minutes. During the mixing, the germanium moves into the acid, and collects in a layer of solution that forms on top of the molten gallium once the stirring stops.

The next step is to siphon off the solution and then concentrate it to produce germanium chloride, which is easily flushed from the solution by argon gas. Once this has been done, the germanium itself can be extracted in a series of reactions to produce a small sample – at most a cubic centimetre – of the gas germane (GeH_4). This is then introduced into a small proportional counter, which detects the decays of any germanium-71 that has been flushed from the reactors. Detecting these decays is the key to counting the solar neutrinos.

There are altogether eight reactor vessels in the SAGE detector, which between them contain 60 tonnes of gallium. Each is fitted with motorised stirrers to provide the essential mixing at the end of every exposure. SAGE started working initially with 30 tonnes of gallium in four reactors, and after 18 months or so of preliminary testing, data collection began in earnest at the beginning of 1990. In June 1990, the team working on the detector revealed their first results, at the *Neutrino '90* conference held at CERN in Geneva. The results were surprising indeed, for SAGE appeared to have detected no solar neutrinos at all!

The standard solar model predicts a capture rate in a gallium detector of 132 SNU, due largely to the capture of the abundant neutrinos from the basic proton–proton reaction. In SAGE, running with 30 tonnes of gallium, such a capture rate should lead to the detection of a total of 20 decays of germanium-71 in the samples extracted over five months of running time. The team monitored each sample for 60 days – more than five times the half-life of germanium-71 – and they did register some counts, but all the signs suggest that these were due to background from unwanted processes rather than from germanium-71 decays.

If the counts in the proportional counters really were due to the decay of germanium-71, then the rate at which they were registered should reflect the half-life; in other words, in the second period of 11 days there should be half the number of counts there were in the first period, and so on. But the counts detected show a flat distribution over time, as might be expected if they are due to a variety of sources. So what had happened to the solar neutrinos, in particular the low-energy neutrinos from the proton–proton reactions, which alone should have produced more than 10 counts in five months? The exciting possibility was that the answer lay with the MSW effect.

Hans Bethe, from Cornell University (who in 1938 had proposed the carbon cycle, one of the basic mechanisms by which stars convert hydrogen to helium) had been one of the first to draw attention to the MSW effect, when in March 1986 he published a paper in *Physical Review Letters* entitled 'Possible explanation of the solar-neutrino problem'. Bethe pointed out that in one solution of the MSW theory, electron-neutrinos above a certain energy would always change identity, and he used the proportion of neutrinos detected in the chlorine experiment, as a fraction of the number predicted by the standard solar model, to calculate the critical energy. He then predicted that the reduction in the gallium experiment would be small, in the region of 10 per cent, as this would detect neutrinos mainly below the critical energy. Shortly afterwards, Peter Rosen and J.M. Gelb and the Los Alamos National Laboratory, produced an argument in support of a different solution to the MSW theory. In this case, the MSW effect would be most marked for the lower-energy neutrinos, and Rosen and Gelb calculated that it could reduce the number of neutrinos detected in gallium by as much as 90 per cent compared with the predictions of the standard solar model.

The first results from Kamiokande II dealt a fatal blow to Bethe's solution, for although this experiment could detect only relatively high-energy solar neutrinos, it detected if anything more neutrinos than the chlorine experiment. Then in June 1990, Bethe together with John Bahcall submitted a joint paper to *Physical Review Letters*, entitled 'A solution of the solar neutrino problem' – dropping the word 'possible' for the first time. They observed that the chlorine experiment, which captures neutrinos with energies greater than 0.814 MeV, detects less than a third of the number predicted; the Kamiokande experiment, on the other hand, by now with an energy threshold of 7.5 MeV, detects about half the predicted number of neutrino reactions. It appears that whatever happens to the neutrinos does indeed have a more severe effect, the lower the energy of the neutrino.

Bahcall and Bethe therefore turned to the second type of solution for the MSW theory and found that by fixing a parameter within the theory, so that the expected number of neutrinos in the chlorine experiment equals the observed number, they could predict the result in Kamiokande II remarkably well. And this solution of the MSW effect would result in the conversion of almost all the low-energy electron-neutrinos to a different type. The rate of 132 SNU predicted by the standard solar model for a gallium detector would be reduced dramatically to 5 SNU – a level that SAGE could confirm only with far more data.

These conclusions were tantalising to the extreme. If correct, they would not only be the solution to a 20-year old puzzle, but would open the door beyond the Standard Model of particle physics. The MSW effect can occur only if the underlying neutrino states have some difference in mass. Bahcall and Bethe found that if they assumed that the degree of mixing between the neutrino states is similar to that observed between quark states, then their solution corresponds to a difference in the mass-squared of the neutrino states of only 2×10^{-7} eV^2. Was this the first positive evidence that neutrinos are not simple massless particles?

An answer in the affirmative would require sound corroborative evidence. However, this time there is no need to wait for 20 years for results from another experiment, as Davis did. A second gallium detector was already nearing completion in 1990, this time under the mountains in central Italy.

In May 1978, Bahcall, Davis and several colleagues had published a proposal, based on Kuzmin's idea, for a new solar neutrino detector that would use 50 tonnes of gallium. Davis had then teamed up with Till Kirsten from the Max Planck Institute for Nuclear Physics at Heidelberg in Germany, and together they led a fruitful collaboration of physicists, making many successful tests on a prototype detector containing 1.3 tonnes of gallium. They proved conclusively the feasibility of the gallium detector and proceeded to propose building a full-scale detector, which would be a joint American–German project. But they failed to persuade the funding authorities in the US to provide the necessary money.

However, Till Kirsten kept the idea alive in Germany, and succeeded in obtaining money for 30 tonnes of gallium (about 20 million DM, or roughly half the overall cost of the experiment) from the West German authorities. He also found a suitable underground location in Italy, some 150 kilometres north-east of Rome, where the mountains of the Gran Sasso Massif rise to more than 2900 metres, forming part of the chain that runs the length of Italy. The Italian authorities had decided to build a new road tunnel to pierce the mountains in this region, and in 1981, Antonio Zichichi, president of the Istituto Nazionale di Fisica Nucleare (INFN), proposed taking advantage of the excavations, by extending them to provide an underground laboratory beside the autoroute. At an altitude of around 1200 m, the laboratory would be well shielded by some 1400 m of rock above it, and would provide a 'quiet' environment for a variety of sensitive experiments.

The laboratory is now built, and one of the major detectors there is based on 30 tonnes of gallium. This is GALLEX, for 'gallium experiment', begun in 1987 by an international team of scientists that Kirsten had drawn together from Grenoble, Heidelberg, Karlsruhe, Milan, Nice, Rome and Saclay in Europe, Rehovot in Israel, and the Brookhaven Laboratory in the US. In many ways, GALLEX resembles Davis's pioneering chlorine detector. The gallium, for instance, is used in a solution of gallium chloride $(GaCl_3)$ and hydrochloric acid; and germanium chloride $(GeCl_4)$ – the molecule made when the gallium captures a neutrino – is swept out by flushing nitrogen through the liquid; lastly, the sample of extracted germanium, once purified, is monitored with a small proportional counter to detect its decays through the process of electron capture.

The 30 tonnes of gallium are contained in 100 tonnes of gallium chloride dissolved in water to simplify the extraction of the germanium, which if produced forms germanium chloride. The germanium chloride is flushed from the liquid by nitrogen, and thereafter the procedure is similar to that with SAGE. The germanium is extracted from the germanium chloride in a series of reactions and converted into germane, which can be introduced into the proportional counter. GALLEX began taking data in 1991, and at the time of writing (January 1992) was poised to show whether the first indications from SAGE really were correct.

Will these experiments solve the puzzle of the missing solar neutrinos? They are certain to guide physicists in the right direction, but if we are to know whether solar neutrinos really do behave as predicted by the MSW effect, then another fundamental piece of information will prove vital – the measurement of the energy of the detected neutrinos. This will be

The GALLEX solar neutrino detector at the Gran Sasso Laboratory in Italy consists mainly of a large tank containing around 30 tonnes of gallium in 55 cubic metres of highly concentrated gallium chloride solution. The photograph shows the arrival of the 9.5-metres long tank at the underground laboratory. (T. Kirsten/GALLEX.)

particularly important as the MSW effect can greatly distort the energy spectrum of the electron-neutrinos, by converting only those at certain energies to another type of neutrino. The dream of many physicists is to measure the energy spectrum of neutrinos from the Sun with sufficient accuracy to distinguish not only between theories but also between their different variations.

Light water, heavy water

> Recently, the possibility of observing ^{8}B-decay solar neutrinos in a large heavy-water Čerenkov detector . . . was raised. That this experiment can be seriously considered is a result of the successful operation of large light-water Čerenkov detectors built deep underground to search for proton decay.[20]
>
> *Herbert H. Chen, 1985.*

It is quite a tall order to ask that you not only detect something as elusive as a low-energy neutrino, but also measure its energy. However, such challenges are exactly what excite the committed neutrino enthusiasts. There is also other information about the solar neutrinos arriving at Earth that cannot be recorded in the chlorine or gallium experiments. At exactly what time do the neutrinos arrive? What direction do they come from? And is it possible to detect the neutrinos if they really have changed from electrons-neutrinos to another type on their way out of the Sun?

Several teams across the world have proposals for detectors that could provide the answers to some or even all of these questions. The feasibility is already being demonstrated in the Kamiokande II detector in Japan, which has shown that it does indeed detect neutrinos that come from the general direction of the Sun.

Kamiokande II contains 3000 tonnes of water, but only the central 680 tonnes can be used in detecting solar neutrinos; the outer region of the detector picks up too many gamma-rays and neutrons from the surrounding rock. This means that the rate of interactions is low, typically one neutrino every six or seven days. But the Kamiokande team have won approval to

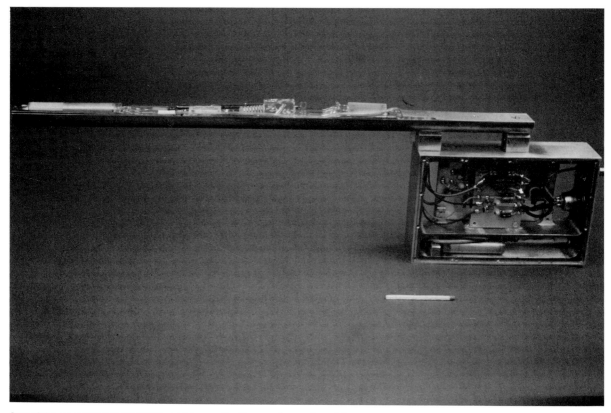

One of the proportional counters used to monitor the small samples of the gas germane extracted from the GALLEX solar neutrino detector, and to detect any decays of germanium-71 back to gallium-71. Such decays should indicate the production of germanium-71 in the detector by solar neutrinos. (T. Kirsten/GALLEX.)

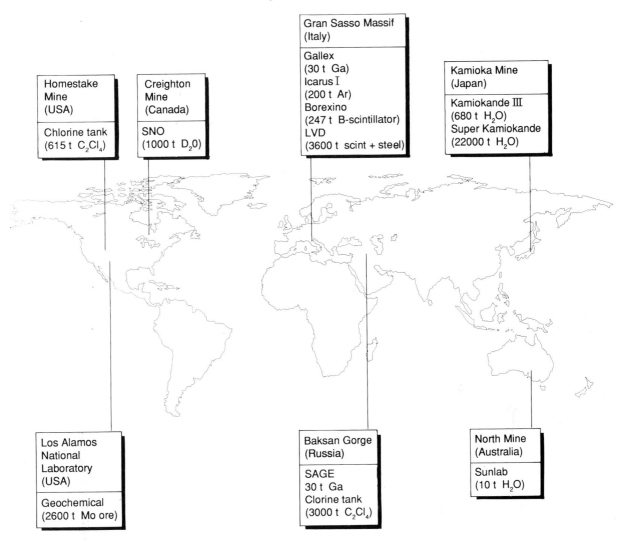

There are now a number of solar neutrino detectors around the world, in various stages from proposals to those that are up and running. This map shows the major detectors that are already built, or being built, at least in prototype form. It indicates the approximate size of the detector in tonnes (t) and the material used to capture the neutrinos. (Ga = gallium; C_2Cl_4 = perchloroethylene; H_2O = water; Ar = argon, B = boron; scint = scintillator; Mo = molybdenum; D_2O = heavy water.)

scale up the whole experiment and build a Super Kamiokande. This will contain 22 000 tonnes of water in which solar neutrinos can be detected and will have double the capacity of Kamiokande II for collecting light. Such a detector, the team claims, would be a thermometer that could measure changes as small as 1 per cent in the temperature of the Sun's core on the time scale of a week. It would also be a superb observatory for cosmic neutrinos (see Chapter 7). Meanwhile, Kamiokande II has been upgraded to Kamiokande III, with new electronics that are prototypes for Super Kamiokande.

Perhaps the most impressive water detector, however, is to be built in Canada, 2070 metres below ground in the Creighton Mine at Sudbury in Ontario, by a team from Canada, the US and the UK led by George Ewan from Queen's University in Ontario. Arguably the most sensitive solar neutrino detector being built, the Sudbury Neutrino Observatory is to be based not on ordinary water, but on 'heavy' water – water in which hydrogen nuclei are replaced by deuterium nuclei, each consisting of a neutron bound with the usual single proton of hydrogen. The idea to build the detector had been largely the brain-child of Herbert Chen, from the University of California, Irvine. He had instigated the discussions with the Canadians, who produce large quantities of heavy water for their nuclear reactors, but sadly he died before the project was approved.

Unlike the chlorine or gallium detectors, which can detect neutrinos in only one way, a heavy water detector can reveal neutrinos through three different basic reactions. The most probable reaction is for a deuteron (a deuterium nucleus) to absorb a neutrino, so that the neutron turns into a proton, just as in the chlorine and gallium experiments. But in this case attention will focus on the electron that is emitted by the reaction. The electron will carry away much of the neutrino's original energy, and will be detected through its tell-tale cone of light. As with Kamiokande, the detector

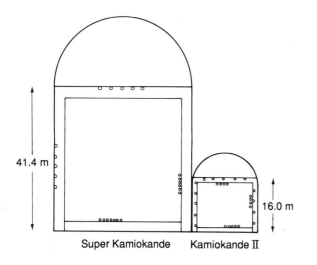

Super Kamiokande   Kamiokande II

41.4 m

16.0 m

The proposed Super Kamiokande detector will have a sensitive volume some 30 times as large as that of the existing Kamiokande II detector. It should detect more than 8000 solar neutrinos a year – enough to observe diurnal or seasonal variations as small as a few per cent.

Georg Ewan (front right), one of the leaders of the project to build the Sudbury Neutrino Observatory, with John Bahcall (front centre) and colleagues at an early planning session at Queen's University, in the summer of 1987. (J. N. Bahcall.)

will only reveal electrons with energies above about 5 MeV at the lowest, and so will trap only neutrinos from boron-8. But with a kilotonne of heavy water, it could be sensitive enough to record boron-8 neutrinos at only 1 per cent of the level predicted by the standard solar model, which would produce an event every three or so days. Moreover, the detector will reveal the energy spectrum of these neutrinos, and so could confirm the MSW hypothesis in which only electron-neutrinos of certain energies change into other types on their journey through the Sun.

The detector will also reveal neutrinos when they simply knock electrons into motion, as in Kamiokande, although these reactions will be only a tenth as likely as the absorptions. However, it is a third type of interaction that makes heavy water so special as a solar neutrino detector. This is a reaction that occurs only through the neutral current (the exchange of a Z^0), and so is equally likely for all three types of neutrino. This means that this reaction should still reveal neutrinos even if they have changed type.

The all-important reaction splits the deuteron into a proton and neutron, but without changing either of them; the neutrino simply gives up some energy and then carries on otherwise unperturbed. Rather as in the experiment of Cowan and Reines, the key to observing this reaction lies with the neutrons. Any nucleus that captures a neutron freed by a neutrino interaction will lose its excess energy by emitting gamma-rays, which will in turn produce tell-tale light signals.

If the boron-8 neutrinos really do change type before they reach the detector, then this deuterium disintegration should still occur at the rate that the standard solar model predicts, for it is blind to the type of neutrino. But if the disintegrations occur at a reduced rate, in line with the results from

Davis's experiment, then it will begin to look as though something really is wrong with the Sun.

The task facing the researchers building the Sudbury Neutrino Observatory is daunting. Having chosen one of the deepest mines in the Western world, the team's main battle will not be against cosmic rays so much as natural radioactivity – particularly when it comes to detecting the neutron produced when a deuteron disintegrates. Unlike Cowan and Reines, they have no other signal to detect in delayed coincidence to differentiate the wanted neutrons from the unwanted ones. Despite the enormity of the task, the potential rewards are clear, as Steven Weinberg has emphasised:

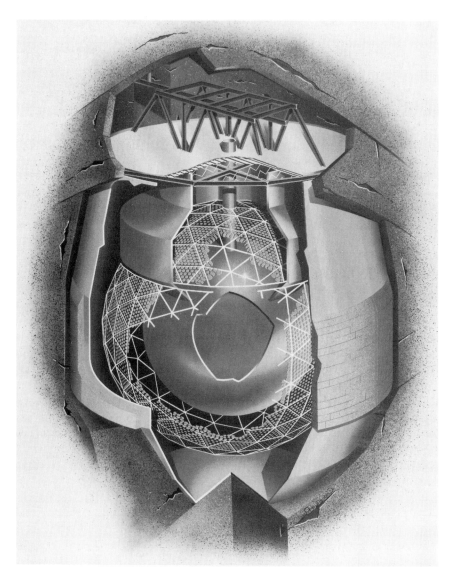

The Sudbury Neutrino Observatory (SNO) is being built in a cavern 30 metres high, excavated 2070 metres underground at a mine in Ontario. It will contain 1000 tonnes of heavy water, held in the spherical vessel of transparent acrylic visible at the centre of this illustration. The heavy water will be surrounded by an outer layer of 7300 tonnes of ordinary water, and will be viewed by the 6400 phototubes supported on the geodesic structure. (Davis Earle, Chalk River Laboratories, AECL Research, Canada.)

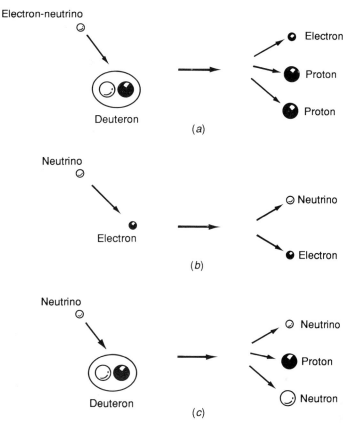

The three reactions through which SNO will detect neutrinos. (a) Inverse beta-decay, in which the neutrino is absorbed, occurs only for electron-neutrinos, but by measuring the energy of the electron produced, SNO will provide an energy spectrum for the neutrinos. (b) Scattering between electrons and neutrinos will also occur, but is more likely for electron-neutrinos than the other types. (c) Neutral current reactions between neutrinos and the deuteron (the nucleus of heavy hydrogen) will occur whatever the type of the neutrino and will provide an important test of the MSW theory of the missing solar neutrinos.

. . . it has now become terribly important to do experiments looking for solar neutrinos, in which one looks . . . for neutral current reactions.

. . . there is a great deal more at stake here than just the interesting question of how the Sun works. Neutrino oscillation whether resonant or not requires a neutrino mass difference to begin with, and the confirmation of a neutrino mass would illuminate some of the deepest questions of particle physics.[21]

The perfect detector?

This experiment just has to be done[22]

Norman Booth, 1980.

All the experiments to measure neutrino energies that I have described so far are sensitive only to the high-energy neutrinos from boron-8. The *ideal* solar neutrino experiment would be one that can measure the energy of neutrinos from the basic proton–proton interaction that in a sense ignites the solar furnace. One such experiment, based on indium-115, was proposed in 1976 by R.S. Raghavan at Bell Laboratories.

When indium-115 (49 protons and 66 neutrons) absorbs a neutrino it changes into tin-115 (50 protons and 65 neutrons). The tin-115 is in an excited state when formed, and decays back to its ground state after about 3 microseconds, by emitting two gamma-rays. The detection of the electron emitted when the indium changes to tin, followed after three microseconds by two gamma-rays of the appropriate energy, signals the absorption of a neutrino. But the key feature is that this reaction occurs with neutrinos with energies as low as 0.12 MeV, which is well below the maximum energy of 0.42 MeV for the neutrinos emitted by the initial proton–proton reactions in the Sun.

So why has no one yet started to build an indium solar neutrino detector? It turns out that there are even more difficulties to overcome than usual, for indium-115 is naturally radioactive. Its half-life is long – nearly half a million billion years (4×10^{14} years). But with 5×10^{24} nuclei in a kilogram of indium, this implies that the rate of natural radioactive decays is a hundred billion (10^{11}) times greater than the rate of solar neutrino absorptions predicted by the standard solar model. Moreover, the majority of the electrons from the beta-decay of indium have a similar energy to those from the absorption.

However, all is not lost. The gamma-rays emitted by the excited tin nucleus after three microseconds provide a unique means of identifying electrons from neutrino absorption, using the same kind of delayed coincidence technique that Cowan and Reines employed. The gamma-rays have well-defined energies of 116 keV and 497 keV. In addition, the one with lower energy will travel less than a millimetre in indium before it deposits its energy, and so its signal should appear close in space to the one detected for the electron emitted 3 microseconds earlier.

How do you detect the electron and the gamma-rays? One proposal was to use liquid scintillator loaded with indium. But initial tests indicated that it would be virtually impossible to detect the low-energy proton–proton neutrinos above the background from the natural beta-decay and gamma-rays from other sources. An alternative possibility, pursued in the early 1980s by Norman Booth and colleagues at Oxford University, is to take advantage of indium's behaviour as a superconductor. Indium is a metal that becomes superconducting – loses all its electrical resistance – at very low temperatures, lower than 3.4 degrees above absolute zero (3.4 K). When this happens, the electrons that take part in conduction become loosely bound together in pairs, which can move easily around the lattice of atoms that forms the metal. An energetic electron or gamma-ray, passing through a piece of superconducting indium, breaks up these pairs, releasing the electrons. The aim in an indium neutrino detector would then be to detect these electrons in some way, the size of the signal being related to the energy deposited by the electron or gamma-ray.

More recently, however, the team at Oxford has turned to making detectors of the compound indium antimonide, which is easier to handle than pure, soft indium metal. In this case, they detect electrons and gamma-rays through the *phonons* they produce – these are vibrations of the basic lattice structure of the atoms in the crystal. The phonons are still detected with a superconductor, but this time with aluminium deposited on the surface of the indium. The initial results from crystals containing several grammes of indium antimonide are very promising. Booth and his colleagues have detected 60 keV gamma-rays with high efficiency, measuring the energy to within about 1 keV.

Interestingly enough, the neutrinos produced by beryllium-7 in the Sun are expected to have an energy spread of about 1 keV, due to thermal motion in the solar interior. Beryllium-7 is made in 15 per cent of the reaction chains leading to helium in the Sun. Some of the time, it captures a proton to make boron-8, which produces the relatively high-energy neutrinos seen in the chlorine and water detectors. But nearly 99 per cent of the time, beryllium-7 captures an electron, and converts to lithium-7, while emitting a neutrino. Because only two objects share the energy released – the lithium nucleus and the neutrino – these neutrinos have a unique energy, rather than a spread in energies (see figure on p.169). This energy of 0.86 MeV is too low to register in the chlorine and water detectors, but it is high enough to produce a reasonable signal in an indium detector. Booth's indium antimonide detector could measure the spread in this energy and provide a direct measure of the temperature at the heart of the Sun, independent of the assumptions of any solar model. One way or another, it seems that a future indium solar neutrino detector could provide us with vital information about the nearest star.

7

Cosmic spaceships

Imagine then the joy and amazement of the world astronomical community when the announcement went out on 24 February 1987 that a bright Type II supernova, clearly visible to the unaided eye, was occurring in a galaxy, 'just next door' . . .[1]

Stan Woosley and Mark Phillips, May 1988.

TEN MILLION OR SO YEARS AGO, amidst the hot, active gases of a small galaxy, a star was born. From the start it was doomed to die in a spectacular explosion, while giant in size but still a mere youngster in comparison to our Sun. In the blink of an eye, the exploding star, or supernova, was to emit as much energy as would 100 Suns in a lifetime of 10 billion years. And 170 000 years later, a miniscule fraction of that energy would be absorbed on Earth in two large tanks of water buried deep underground.

The detection of this energy, on 23 February, 1987, betrayed the death of the star, but it also marked the birth of a new branch of observational astronomy here on Earth – the beginnings of neutrino astronomy. For the first time, experiments had detected neutrinos from beyond our own Galaxy.

Some three hours after the burst of neutrinos swept through our planet, visible light from the supernova began to arrive at Earth. It was noticed first, nearly a day later, by Ian Shelton, an astronomer at the Las Campanas Observatory in Chile. He was photographing the Large Magellanic Cloud, one of two small galaxies that orbit our own Milky Way galaxy. Shelton had planned a routine search for variable stars and novae – faint stars that show a sudden brightening. On his third night of observing, 24 February, he struck lucky. On developing the photographic plate he had exposed for three hours at the 25-centimetre refracting telescope, he noticed a new star. Then, as Shelton's colleague Robert Jedrzejewski recalled:

He walked into the control room and asked what magnitude a nova should shine in the Large Magellanic Cloud. [We decided] approximately 8th magnitude.

Ian then inquired about the identification of an object if its apparent magnitude was $+5$ while on a plate on the previous night it had been about $+12$ or fainter. Barry

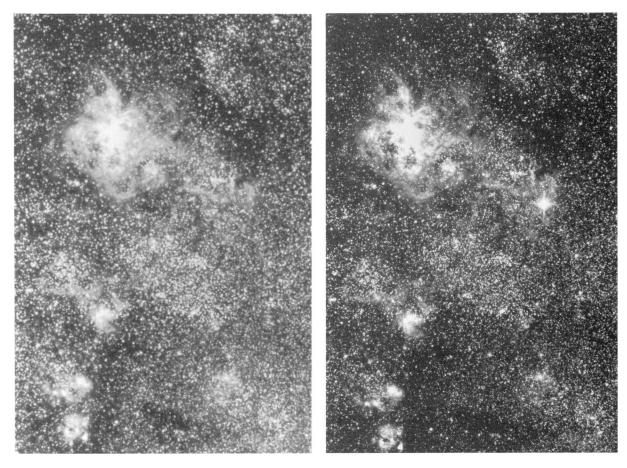

Supernova 1987A erupted in the region of the Tarantula Nebula, on the edge of the Large Magellanic Cloud, on 23 February, 1987. These images show the region before the explosion (left) and four days after, on 27 February. The supernova appears clearly to the lower right of the Tarantula Nebula, which glows red due to emission of light from hydrogen atoms. (European Southern Observatory.)

[Madore]'s answer was that it must be a supernova. At this point Oscar [Duhalde] chipped in and confirmed that he had sighted the object visually earlier in the evening. We then all walked out of the dome to see for ourselves. The night was exceptionally transparent and the new . . . star was easily visible.[2]

Duhalde, the night assistant, had noticed the star a couple of hours earlier, when he had gone out to look at the sky, but for some reason he neglected to mention it to the observers. 'We must have been working [him] too hard',[3] Jedrzejewski later remarked.

The discovery was soon confirmed. Shelton's exposure that revealed the supernova had ended at 4:20 Universal Time (UT) on 24 February. On the previous night in New Zealand, an amateur astronomer, Albert Jones, had been observing the same part of the sky through his small telescope and had noticed nothing unusual; however, on the night of 24 February, when he looked again at the sky, he saw a new star close to the Tarantula Nebula, with a magnitude he estimated to be between 5.0 and 7.0. Jones had independently discovered the supernova less than 4 hours after Shelton had noticed it on his photograph.

Thus did the combined ranks of amateur and professional astronomers mark the beginnings of one of the most significant astronomical events of the twentieth century – the emergence of SN1987A, the first supernova of 1987, and the first to be clearly visible to the naked eye for nearly 400 years.

The event of a lifetime

> The collapse and explosion of a massive star is one of nature's grandest spectacles. For sheer power nothing can match it. During the supernova's first 10 seconds . . . it radiates as much energy from a central region 20 miles across as all the other stars and galaxies in the rest of the visible Universe combined . . . It is a feat that stretches even the well-stretched minds of astronomers.[4]
>
> *Stan Woosley and Tom Weaver, August 1989.*

Astronomers soon measured the location of the new supernova accurately, and searched their charts for what had lain there previously. The star that had exploded, and provided by far the brightest supernova since the invention of the telescopes, was evidently Sanduleak − 69°202, named for its position in the sky and Nicholas Sanduleak who catalogued the star in 1969. The star had been a 'blue supergiant', a star with some 20 times the mass of the Sun and 50 times the radius, and a surface temperature (revealed by its blue colour) of around 12 000 degrees, twice that of the Sun (which appears yellow).

The identification was something of a surprise to many astronomers, who had quickly identified SN1987A as a Type II supernova, or in other words, a supernova whose light emissions indicate the presence of hydrogen. (Type I supernovae show no evidence for hydrogen.) Previous identifications had always shown that Type II supernovae form from *red* supergiants. The standard theory was that when such a massive star runs out of fuel at the centre, gravity makes its core collapse, triggering a shock wave that blows apart the outer layers. But there had been no observational evidence to suggest that a blue supergiant could lead to a Type II supernova, although a few people had found that it could happen in computer models. Sanduleak − 69°202 provided the first glorious example to be observed.

Over the past 20 years, a number of supernova specialists, in particular Tom Weaver at the Livermore Laboratory in California, Stan Woosley at the University of California (Santa Cruz) David Arnett at the University of Arizona, and Ken'ichi Nomoto at the University of Tokyo, have developed computer models to simulate what happens inside massive stars. The simulations show that these stars evolve to become something like gargantuan onions.

Take the case of Sanduleak − 69°202. According to Weaver and Woosley, whose model reproduces the observed rise and subsequent fall in brightness of SN1987A, the progenitor star was born about 11 million years ago in the Tarantula Nebula, a bright spider-shaped region of hot, glowing gas in the Large Magellanic Cloud. Like the Sun, Sanduleak − 69°202 began life composed mainly of hydrogen gas, and was initially fuelled by 'burning' hydrogen to form helium. The nuclear energy released in this way held off

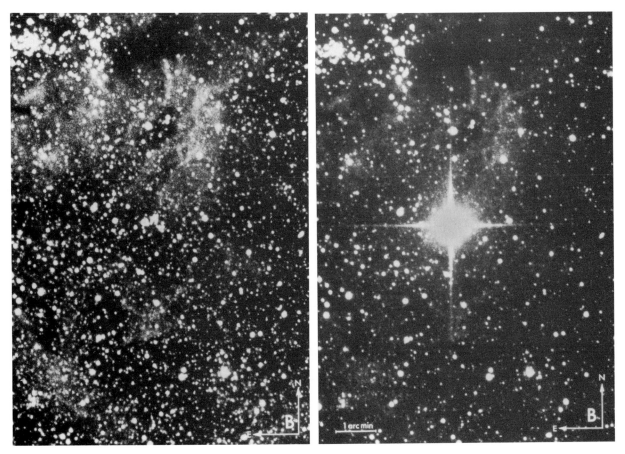

The star that had exploded as SN1987A was soon identified as a bright blue star, Sanduleak −69°202, which appears as the leftmost of the two bright spots near the centre of the image on the left. The image on the right, taken on 26 February, shows the supernova three days after it appeared (the cross is an optical effect in the telescope). Both images were recorded in blue light, the one before the super- nova taking 60 minutes, the one after only 15 minutes. (European Southern Observatory.)

the gravitational urge to contract. The star existed like this for 10 million years, until all the hydrogen in the central core was completely converted to helium.

With its reserve of nuclear energy temporarily depleted, the core began to shrink and heat up, making the outer layers of the star expand. Within 50 000 years or so, the core had become hot enough and dense enough for helium to begin to fuse together to form carbon and oxygen, while hydrogen in the layer around the core continued to make helium. Meanwhile, the star had swollen from a radius of about 4 million kilometres to one of 300 million kilometres, more than 500 times the radius of the Sun. Sanduleak −69°202 was now a red supergiant.

For nearly a million years, the helium at the star's core continued to burn, until that too was gone and gravitational contraction set in once again. This time once the core had become sufficiently hot and dense the carbon began to burn, forming nuclei of neon, sodium and magnesium. By now, the star had begun to take on its 'onion-skin' appearance, with a layer of helium surrounding the core of carbon, all wrapped up in an outer envelope of

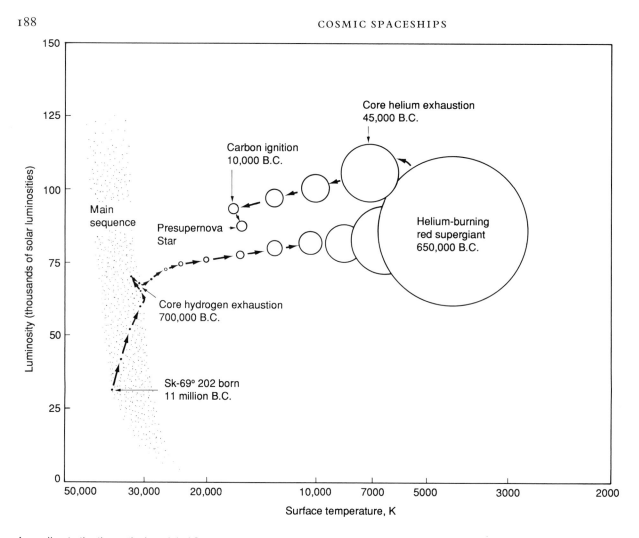

According to the theoretical model of Stan Woosley and his colleagues, Sanduleak −69°202 was probably born some 11 million years ago, with a mass about 18 times that of the Sun. Its initial size predetermined its future life, which is mapped out on this diagram showing the luminosity versus surface temperature at various stages, until the moment immediately before the supernova explosion. Notice how, once the star had burned all the hydrogen at its centre, its outer layers expanded and cooled until it became a red supergiant, over to the right on the diagram. At that stage helium started burning in the core to form carbon, and by the time the supply of helium at the centre was exhausted, the envelope contracted and the star became smaller and hotter, turning into a blue supergiant. (Courtesy Tom Weaver, Stan Woosley and John Maduell, Lawrence Livermore National Laboratory, Livermore, California.)

hydrogen. At this stage the star lost some of its huge envelope – a shell of gas detected some 40 000 years later when ionised by the flash of ultraviolet light from the supernova explosion itself. Moreover, the temperature evidently did not stay high enough to sustain the remaining envelope, which began to contract, and as the star became smaller it turned from red to blue. It became a blue supergiant.

Sanduleak − 69°202 was now well on its roller-coaster ride to destruction. Carbon burning took only about 12 000 years, to be followed by neon burning and oxygen burning, each for only a few years. The reason for this rapid progression lies, perhaps surprisingly, with neutrinos. When the core reached a temperature of 500 million degrees, during carbon burning, the photons it radiated were sufficiently energetic to be able to produce pairs of electrons and positrons. These particle–antiparticle pairs would usually annihilate back to photons (gamma-rays), but sometimes they would produce neutrino–antineutrino pairs. And the feebly interacting neutrinos and antineutrinos could readily escape from the core and from the star itself, taking with them energy that would otherwise have been expended in holding off gravitational collapse. In the words of Woosley and Weaver:

As the core's temperature rose during the later stages of its evolution, the neutrino luminosity rose exponentially to become a ruinous energy drain, hastening the star's demise.[5]

Finally, silicon and sulphur at the core, the products of oxygen burning, burned to produce iron, in only a week or so, and then Sanduleak − 69°202 was poised for collapse. Iron nuclei cannot release energy on fusing together; indeed the process requires energy. So, the furnace at the centre of the star simply went out, and there was no longer anything to slow down the onslaught of gravitational collapse. Once the iron core reached a critical mass of 1.4 times the Sun's mass, when its diameter was about half that of the Earth, the star's fate was sealed.

Within a few tenths of a second this dense iron ball had collapsed to something only 50 kilometres across, velocities in the outer part reaching a quarter the speed of light. But then the collapse of the inner part halted as it became denser than an atomic nucleus, and the nuclear force between protons and neutrons began to prevent them from being squeezed any closer together. Not only did the inner region halt in its contraction, but it sprang back and collided head on with the infalling outer layers. The result was a tremendous shock wave that set off back through the outer region of the core.

It is truly remarkable that it can all happen so swiftly. As Adam Burrows, a theoretical astrophysicist at the University of Arizona, has written:

The star may have lived for 10^7 years, but its core collapses in a second. Within a day, the entire star is disassembled.[6]

Neutrinos and the weak force that creates them played a vital part in bringing the star to its sudden death. Theoretical models suggest that they also have a critical role in the ensuing explosion. Strangely enough, it is probable that

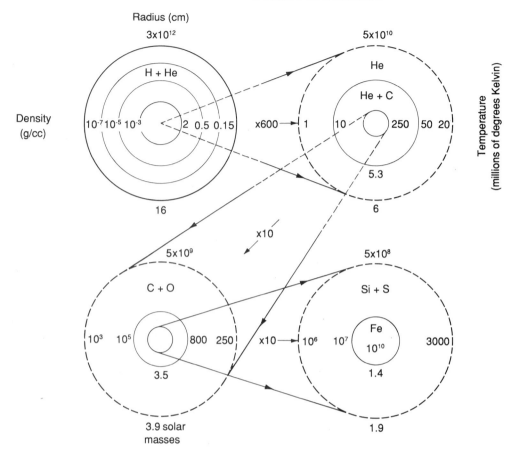

Radius (cm)

Density (g/cc)

Temperature (millions of degrees Kelvin)

neutrinos nearly stopped SN1987A from ever happening, but that they also got it going again.

As the shock wave travelled towards the edge of the core it would lose energy, in part through the creation of neutrino–antineutrino pairs, which could escape readily from the outer regions of the core. And on reaching the edge of the core the shock wave probably stalled, a large proportion of its energy taken away by the elusive neutrinos.

However, the inner core would be very hot, hot enough indeed to produce very energetic gamma-rays that, in turn, make more neutrino–antineutrino pairs. And the density of the inner core would be high enough for the neutrinos to take several seconds to diffuse out, taking thousands of times longer than if the density were lower. These neutrino interactions may have been just what was needed. James Wilson and Ron Mayle at the Livermore Laboratory have calculated that if a few per cent of the neutrinos dumped their energy in material behind the stalled shock, but close to it, then this could be sufficient to set the shock wave going again, so that SN1987A's optical firework display could begin.

Immediately before its collapse, Sanduleak −69°202 would have resembled a giant stellar 'onion', made up of concentric layers of different elements. This diagram shows the structure calculated by Stan Woosley and his colleagues, with the centre magnified on three successive enlargements by factors of 600, 10 and 10 respectively. The radius of the area shown is given in centimetres at the top of each diagram, the corresponding mass (in solar masses) at the bottom; the figures to the left of the core of each diagram show the density of different layers (in grams per cubic centimetre) while the numbers to the right give the temperature in millions of degrees Kelvin.

SN1987A offered astronomers their first opportunity to harness an armoury of modern technology in the study of a supernova. To quote Woosley and Weaver:

For us and hundreds of others, theorists and observers at all wavelengths collaborating to document and explain one of the heavens' grandest events, it has been a time of matchless exhilaration, scientific cooperation and intellectual reward – the event of a lifetime.[7]

Nor was it only the astronomers and astrophysicists who shared in the excitement.

The first messengers

> ...the crowing first associated with the LMC supernova must be the independent epochal detections ... of the brief ($\sim$ 10 second), but prodigious, neutrino burst from its core. For the first time, we have penetrated the otherwise opaque supernova ejectum and glimpsed at the violent convulsions that attend stellar collapse.[8]
>
> *Adam Burrows, 1988.*

As long ago as 1932, within a few months of Chadwick's discovery of the neutron, the Russian physicist Lev Landau predicted the existence of neutron stars. These are stars in which matter is so greatly compressed that the protons and electrons within it have coalesced to form neutrons. Within two years, Walter Baade and Fritz Zwicky in California had made the connection to supernovae. They proposed that the enormous energy associated with a supernova could result from the collapse of an ordinary star to a neutron star. In such a collapse, the core of the star – all one and a half solar masses worth – becomes squeezed into an extremely dense object only 10–20 kilometres across. For this to happen the core must lose an unimaginably vast amount of energy – the so-called gravitational binding energy. The amount is relatively simple to calculate, and comes out at $(2–3) \times 10^{53}$ ergs. To put this in context, it is about 100 times the total output of the Sun during its expected lifetime of 10 billion years. And we know that 99 per cent of the energy from the stellar collapse comes out not in the optical firework display, which accounts for less than 1 per cent, but in the form of neutrinos which stream outwards in all directions through intergalactic space.

At the distance of the Earth, 170 000 light years from the source of the explosion, the swarm of neutrinos from SN1987A would have thinned considerably, although there would still be a phenomenal 50 billion per square centimetre to stream through our small planet. Most passed us by, as impervious to our presence as we were to them. They raced through the Earth on its southern hemisphere, to exit on the northern side. Of course, some would not make the whole journey through the Earth, and nearly 20, we know, were caught just as they were escaping from the northern hemisphere. For these neutrinos were stopped in underground detectors in Japan and the northern US.

As soon as word of the new supernova began to spread, astrophysicists and particle physicists around the world began to consider the implications for neutrino detection here on Earth. From calculations on the backs of envelopes and beer mats to more detailed computer analyses, a veritable shock wave of computation spread rapidly through the community. Among the first off the mark were John Bahcall and colleagues, whose letter to the journal *Nature* was received on 2 March, only three days after the supernova. They had calculated the numbers of neutrinos that existing experiments might be expected to detect. Davis's solar neutrino detector in the Homestake Mine (see Chapter 6) would probably produce one extra atom of argon, which would be indiscernible against the normal production rate of 1–2 atoms per day. But according to Bahcall *et al*, the Kamiokande II detector in Japan (see also Chapter 6) stood a good chance of detecting up to 50 or so neutrinos, all within a space of a few seconds – the time taken for a supernova to collapse and emit a large proportion of its neutrinos. The appearance of a number of particles so closely spaced in time would clearly stand out against normal background events.

The time to look for such a neutrino burst was a little before the first optical observations of the supernova. Within the burgeoning supernova, the neutrinos, once they had escaped, would move faster than the shockwave and so would be the advance guard for the light emitted as the shock passed through the layers outside the iron core. And when the team working on the Kamiokande detector looked back through the data recorded at the end of February they found such a burst of events. Around 7 : 35 Universal Time on 23 February, 11 events had occurred that looked like interactions due to neutrinos, all clustered within 13 seconds, the first nine of them being within 2 seconds of each other. This was about 18 hours before Ian Shelton first noticed the supernova. The Kamiokande team was lucky. The burst occurred only minutes after a routine calibration of the detector, which would have completely obliterated any signals from neutrinos.

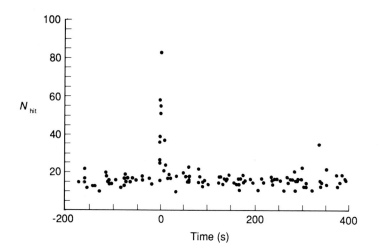

The number of phototube hits recorded in Kamiokande II on 23 February 1987 shows a dramatic increase above the usual background level at about 7:35 Universal Time (Time 0 on this plot). This demonstrates clearly that something unusual happened in the detector around this time, and that the events attributed to SN1987A are not likely to have been a random fluctuation.

Small circles mark the photo-tubes that fired in Kamiokande II in this reconstruction of the highest-energy 'event' detected during the burst of neutrinos from SN1987A. The area of each tiny circle is proportional to the amount of light detected by the phototube at that location. The arrow shows the origin and the direction of the positron pro-duced by the neutrino's interac-tion with a proton in the water, and the additional line traces out the places where the cone of Čerenkov light produced by the positron intercepted the walls of the detector. (Dotted lines indi-cate features on the far surfaces of the detector.) (M. Koshiba, Kamiokande II.)

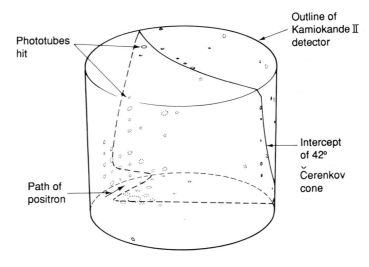

Phototubes hit

Outline of Kamiokande II detector

Intercept of 42° Čerenkov cone

Path of positron

Another team with an underground detector knew they too had a good chance of having observed the neutrinos from SN1987A, and when they heard of the discovery in Kamiokande, they knew where to look in their data. This was the IMB detector, named for the original participating institutions of the University of California, Irvine, the University of Michigan and the Brookhaven National Laboratory. (The team includes two people intro-duced in earlier chapters, Maurice Goldhaber and Fred Reines.)

The detector, a tank of some 8000 tonnes of purified water, resides in a cavern 600 metres below ground in the Morton–Thiokol salt mine near Fairport, Ohio. It was built, like the Kamiokande detector, to search for the possible decay of protons. Light-sensitive phototubes, 20-centimetres in diameter, lie at 1-metre spacings on the six inner surfaces of the tank. They detect Čerenkov radiation, that is, light emitted when electrically charged particles move through the water faster than the speed of light in the water. The light spreads out in a cone around the path of a particle, and strikes a ring of phototubes as it arrives at the side of the detector (see figure on p.162).

Neutrinos normally provide unwanted background signals in the IMB detector; their interactions in the water, which occur about twice a day, can uncannily mimic the decay of a proton. But the IMB team realised in the early 1980s that neutrinos from a supernova might easily be detected. At the time, they considered the effect of supernovae at the centre of our own galaxy. A stellar collapse so nearby would flood the detector with neutrinos, making Čerenkov rings from individual particles impossible to resolve. Instead, the team believed they might detect the sudden increase in signals from all the phototubes. (One member of the team, John LoSecco, sent a paper to *Nature* discussing the possibilities; the staff at *Nature* were apparently not too impressed, as it took them 14 months to publish the paper!)

As it happened, when the opportunity to observe a supernova arose, as

Sanduleak −69°202 blew apart, the event took place far enough away not to swamp the detector with signals, but still close enough to provide a convincing effect. At precisely 7:35:41:37 UT, the first of a bunch of eight neutrinos fired the phototubes of the IMB detector. A further seven triggered the detector within the next six seconds. Unfortunately, at the time a quarter of the phototubes were not working because their power supply had failed, otherwise the detector would have revealed still more neutrinos. But as it was, the negligible background rate of accidental triggers and the simultaneous arrival of neutrino bursts in detectors on separate continents were together conclusive enough evidence to link the signals in IMB and Kamiokande to the supernova.

A paper describing the observation of neutrinos from SN1987A by the Kamiokande team arrived at the offices of *Physical Review Letters* on 10 March, three days before a similar one from the IMB collaboration, and the two papers were published together on 6 April, six weeks after the appearance of the neutrinos from the supernova.

Our events and their proximity in time to the optical observation of the supernova 1987A are compelling evidence that neutrinos have been seen from a supernova collapse.[9]

Thus concluded the IMB team. The researchers on Kamiokande went further:

This . . . is the first direct observation in neutrino astronomy, and coincides remarkably well with the current model of supernova collapse and neutron-star formation.[10]

And Stan Woosley and Tom Weaver could later write:

The theoretical significance of the neutrino detection was considerable . . . the fleeting detection of the neutrino burst shows that, as theory had predicted, a neutron star is formed in a Type II supernova.[11]

When the core of a massive star collapses to form a neutron star it should emit some 10^{57} electron-neutrinos, as protons combine with electrons to form neutrons (a form of inverse beta-decay). These should emerge with an average energy of 10 MeV, a figure that can be calculated from the temperature and density of the core in the computer models of stellar collapse. They should carry a total energy of around 1.3×10^{52} ergs away from the collapsing star, which is only a twentieth of the estimated total energy released.

But in addition there are the neutrino–antineutrino pairs produced in the high-temperature environment of the collapsing star – the particles that do so much in bringing about the star's cataclysmic death. Neutral-current interactions (that is, via the exchange of the Z^0 particle, described in Chapter 5), which were unknown on Earth until the early 1970s, lead to the production of all types of neutrino and antineutrino. In other words, muon- and tau-neutrinos and antineutrinos should be produced in addition to the

Inside the water-filled IMB detector, 600 metres below ground in the Morton–Thiokol salt mine near Cleveland, Ohio. A regular array of 2048 phototubes monitors the water, each tube mounted on a square plastic sheet, designed to collect light that does not strike the tube directly. The diver is making one of the monthly inspections of the detector. The distance to the far wall is 20 metres, demonstrating the extreme clarity of the purified water in the detector. (IMB Collaboration.)

This computer reconstruction shows the ring of Čerenkov light produced in the IMB detector by the interaction of one of the neutrinos from SN1987A. The long straight lines outline the walls of the detector; the short lines and crosses mark the locations of phototubes that detected some light, the number of lines at each point being proportional to the amount of light detected. (IMB Collaboration.)

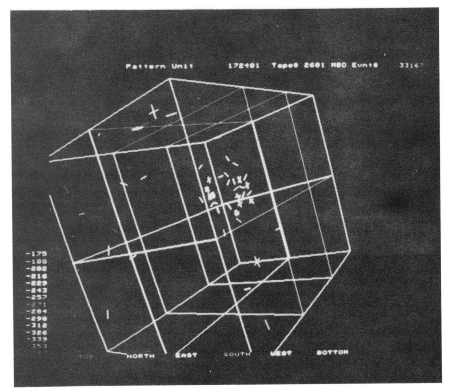

electron species, which can also be produced through charged-current interactions. Moreover, it turns out that the total energy carried away by each type should be more or less equal.

The Kamiokande and IMB detectors provided the first test of these ideas. They can detect all species of neutrino through their collisions with electrons in the water: the electron moves off in roughly the same direction as the incoming neutrino and produces its tell-tale ring of Čerenkov light on the wall of the tank. But electron-antineutrinos can also react with protons, to produce a positron and a neutron, and they do this more readily. (This is the reaction Cowan and Reines studied in the first detection of the neutrino in the 1950s, as Chapter 2 describes.) In this case, the positron, which gives rise to the Čerenkov light, can emerge in any direction.

In the final analysis, only one of the detected particles, in Kamiokande, appears to point back towards the Large Magellanic Cloud, and so could be due to an elastic scatter between a neutrino and an electron. The majority of the events, as expected, appear to be due to the capture of electron-antineutrinos by protons. It is then possible to estimate from the few electron-antineutrinos detected and their average energy (measured via the amount of Čerenkov light from the detected positrons) to estimate the temperature of the supernova's core and the total energy carried away by all types of neutrino and antineutrino.

The results are astonishing, dramatic evidence that our picture of stellar collapse is basically correct. At the *Neutrino '88* conference, held at Tufts University in Massachusetts, Adam Burrows observed that:

Core collapse has been studied for 30 years and neutron stars for 50 years in blissful theoretical isolation. That the theorists were 'on the money' is gratifying . . .

. . . these detections [of neutrinos from SN1987A] provide us with the first definitive tests of the basic theory connecting stellar death, supernovae, and neutron star birth.

Within ∼ 10 seconds on Feb. 23.316 (UT) 1987, that theory was transformed into an astronomy.[12]

The detected neutrinos indicate that SN1987A emitted a total energy of $(2–3) \times 10^{53}$ ergs. This is equal to the calculated gravitational binding energy that would have to be released by the collapse of a core of about 1.5 solar masses to a neutron star. Moreover, the neutrinos arrived at Earth over an interval of several seconds. This confirms that they really did diffuse out from the dense core relatively slowly, as in the computer simulations. The one weak link in the chain, which remains unproven, is the neutrinos' role in the 'revitalisation' of the stalled shock – the mechanism that gets the shock wave going again just when it has run out of energy.

We will not have to wait 400 years for another supernova to provide the answer to these questions. With SN1987A the science of neutrino astronomy has been born, and neutrino astronomers now know what the burst from a supernova looks like. This means that they should be able to recognise a supernova event if its optical signal is too faint to see or is

A plot of the arrival time of neutrinos in both the Kamiokande II and the IMB detectors (with zero equal to 7:35:40 Universal Time on 23 February, 1987) shows how neutrinos continued to arrive for up to 12 seconds. This apparently confirms theoretical ideas that the neutrinos should diffuse out from the dense core of a supernova over a period of several seconds.

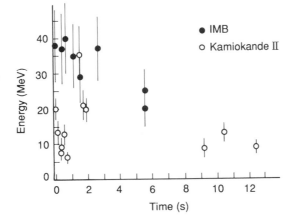

obscured from view, for example by dust in the main disc of our own galaxy. At the *Neutrino '88* conference, Lev Okun from the Institute for Theoretical and Experimental Physics in Moscow made the following comments:

> The main impact of SN1987A is that it served as a spectacular rehearsal. More physicists are thinking now about the heavens than they did before February 24, 1987. The date of the next and the main performance will be announced at the end of this talk.[13]

Okun was not joking. Indeed, he was 'absolutely serious':

> To predict the year of explosion of a supernova is not harder than to predict the year of funding a big accelerator or a big detector.[14]

His line of arguing is quite simple. Only six supernovae have been seen in our galaxy during the past 1000 years and these were all relatively close by, in the region of the Sun, away from the centre of the galaxy. But observations of other galaxies have shown that supernovae occur most often near the centre of galaxies, the very region in our galaxy that is obscured from our view by dust. And as Okun commented:

> . . . that explains why our ancestors have not seen supernovae there. But dust is transparent to neutrinos.
>   According to the most recent estimates of experts there occur in our Galaxy two to three supernovae bangs each century. Big neutrino detectors were built rather late, so we may have missed our 20th century chance. But being optimistic, I expect that the date of the next supernova is 2003 ± 15 years . . .
>   . . . What is especially important . . . is the minus sign. The supernova neutrinos may come to this hall even before I finish my talk. So we have to hurry![15]

Okun need not worry. Inspired by SN1987A, would-be neutrino astronomers around the world are already building new detectors for the 1990s, and dreaming of those that they may build in the next century. The hope is that the new-born science may not have to wait too long to reach maturity.

Extraterrestrial neutrinos

> One may even anticipate eventual high-energy neutrino astronomy, since
> neutrinos travel in straight lines, unlike the usual primary cosmic rays . . .; and the
> neutrinos will convey a type of astronomical information quite different from that
> carried by visible light and radio waves.[16]
>
> *Kenneth Greisen, 1960.*

Ten years after his work with Clyde Cowan, proving that Pauli's neutrino
did indeed exist as a real particle that could interact with matter, Fred Reines
was to be found working some 3200 metres below ground in a gold mine in
South Africa. Once again he was searching for neutrinos, but this time not
'man-made' neutrinos, created in the heart of a nuclear reactor. Instead, he
was looking for high-energy neutrinos produced naturally in the Earth's
atmosphere by energetic particles originating well beyond our Solar System.
The apparatus began working in the autumn of 1964. Then, as Reines and
his colleague J.P.F. Sellschop wrote later:

. . . on February 23, 1965, the detectors recorded a muon that had travelled in a
horizontal direction – the first 'natural' high-energy neutrino had been observed![17]

Neutrinos with energies that range up to values far beyond anything that
we can produce in laboratories stream through us constantly. Those with the
lowest energies are the most common. The neutrinos from the Sun have
energies of 14 MeV at most, and then only when they come from the
relatively infrequent decays of boron-8; those from SN1987A had an
average energy of about the same value, in line with theoretical models of
Type II supernovae. But other neutrinos arrive at the Earth's surface with
energies thousands of times greater and more. They are created in the
interactions of high-energy charged particles with matter – in the Earth's
atmosphere, in interstellar space, and farther afield in other star systems, and
probably in other galaxies. While you read this page, about two or so
neutrinos with energies greater than 10^{10} eV (10 000 MeV), pass through it
each second, having been produced in nuclear reactions in the atmosphere.

The Earth's upper atmosphere is bombarded by a steady hail of very
energetic cosmic ray particles, mainly protons. When the cosmic rays collide
with atomic nuclei in the atmosphere they produce charged and neutral
pions (see Chapter 4); when the charged pions decay, they usually produce
muons and muon-neutrinos; and often, before they reach the Earth's
surface, the muons also decay, to electrons, electron-neutrinos and more
muon-neutrinos. The cosmic-rays that generate these 'showers' of secondary particles can have energies as high as 10^{20} eV or more, 100 million times
greater than the highest energies that man-made particle accelerators
presently achieve.

The cosmic-rays with the highest energies are rare indeed. Particles with
an energy above 10^{16} eV pass through an average room about 100 times a
year, while energies greater than 10^{20} eV are detected at a rate of only a few

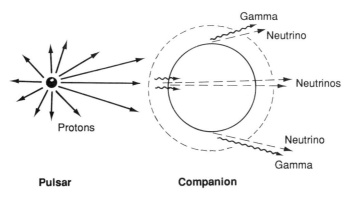

Pulsar

Companion

A possible source for very-high-energy cosmic-rays might be a binary star system consisting of a pulsar (a rapidly rotating neutron star) and a companion, with material from the companion being drawn towards the dense neutron star. Particles accelerated in high fields around the pulsar could interact with the material, producing charged particles as well as neutrinos and gamma-rays. The charged particles would be curled away by the magnetic fields in space, to join the mix of cosmic-rays, but neutrinos and gamma-rays would not be deflected and any detected on Earth would reveal the direction of the source. In some cases, the gamma-rays might be lost through absorption in intervening matter, leaving only the neutrinos to be detected.

per square kilometre per century. But these ultra-high-energy cosmic-rays can clearly give birth to neutrinos of very high energies.

The origin of cosmic-rays remains uncertain. Any information about the direction of charged particles from outer space becomes lost as they pass through the galactic magnetic fields, twisting and spiralling, to arrive at Earth almost uniformly from all directions. One suggestion is that charged particles might be accelerated in shock waves, such as a supernova produces. Another idea, which is currently popular, is that cosmic-rays may gain their high energies in the environment of exotic astronomical objects such as pulsars – rapidly rotating neutron stars, which are often the remnants of Type II supernovae.

But gamma-rays are not affected by magnetic fields. Although the number of high-energy gamma-rays in the cosmic radiation is less than 0.1 per cent of the number of charged particles, several research teams have built detectors to search for the showers of electrons, positrons and gamma-rays that a high-energy gamma-ray generates on interacting with a nucleus in the atmosphere.

In recent years these detectors have revealed a number of sources of very-high-energy gamma-rays, with energies well above 10^{12} eV. Gamma-rays with lower energies can be explained away as radiation from accelerating electrons. But it seems difficult to believe that radiating electrons could be the source of the higher-energy gamma-rays. What seems more plausible is that protons, say, are somehow accelerated to these high energies, and that they interact with matter, producing showers of pions. The neutral pions will soon decay to gamma-rays. So perhaps the sources of the very high-energy gamma-rays are the very same cosmic accelerators that whip protons, and to a lesser extent other nuclei, up to energies far higher than we will achieve on Earth in the foreseeable future.

If these sources produce neutral pions, they must also produce charged pions, and these will produce neutrinos as they decay to muons. The muons, being charged will be deflected by the magnetic fields in space, but the neutrinos, like the gamma-rays, will be unaffected and should travel directly

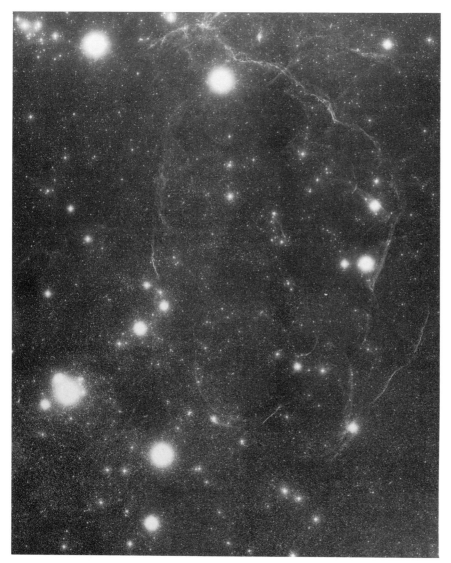

Bright wisps of gas are the visible remains of a supernova that occurred some 11000 years ago in the constellation Vela, visible from the Southern Hemisphere. At the heart of the remnant, near the left of this image but not visible, lies a pulsar. This is a rotating neutron star, which is the collapsed core of the star that exploded, and which is possibly a natural accelerator for cosmic rays. Certainly there is evidence that this region is the source of very-high-energy gamma-rays and so could also emit cosmic neutrinos. (Royal Observatory, Edinburgh.)

to Earth. If a detector were to pin-point a source of ultra-high-energy neutrinos, then it would be the best evidence yet for a cosmic accelerator. For while some question mark hangs over the origin of the gamma-rays, there is little doubt that neutrinos of similarly high energies must come from nuclear interactions between subatomic particles.

So the challenge of neutrino astronomy is still beckoning, 25 years after Reines's pioneering experiments. Now a number of experiments are being set up across the globe, and many more lie on the drawing boards or are crystallising in the minds of physicists pulled by the challenge of chasing something that is nearly impossible to detect.

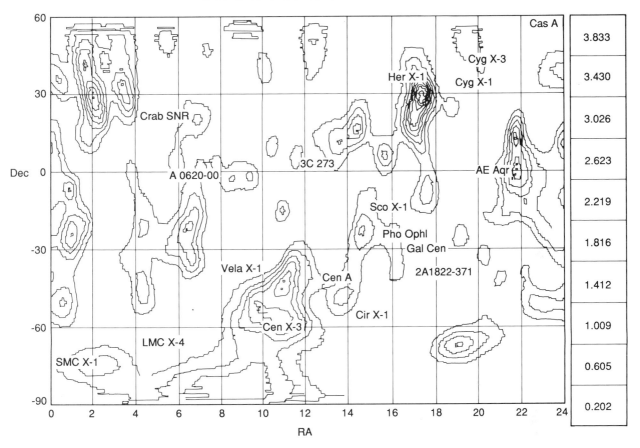

This 'neutrino sky map' is based on the distribution of cosmic neutrinos detected in the IMB water Čerenkov detector (see p.195). It shows the probability that the number of neutrinos detected exceeds a random background. Though not built as a neutrino observatory, IMB serves to provide a tantalising glimpse of things to come with larger detectors with hints of higher numbers of neutrinos in a few places, for example near Hercules X − 1. (Ralph Becker-Szendy, University of Hawaii.)

Neutrino telescopes

> It is immediately clear that the detector must be large. . . . Also, one must be able to avoid, to the extent of about one a year, background events that may simulate neutrino interactions. . . . It is necessary to set up the experiment far underground to reduce the effect of the cosmic radiation other than neutrinos.[18]
>
> *Kenneth Greisen, 1960.*

The 1990s will begin the age of neutrino astronomy. Many teams of researchers are at work on neutrino 'telescopes' which range from ideas on paper, through definite but as yet unapproved proposals, to detectors that are already lying in patient wait for extraterrestrial neutrinos. Some of these telescopes are designed specifically to detect neutrinos from the Sun, as described in Chapter 6; others are geared towards the detection of neutrinos from supernovae, mainly within our own Galaxy; and those in a third group have their sights set farther afield, in search of distant sources of very-high-energy neutrinos, which will perhaps provide the key to the mystery of the origin of cosmic-rays.

These 'neutrino telescopes' bear little resemblance to conventional optical and radio telescopes. To begin with, they are usually located under mountains rather than on top of them. This is to shield them from the majority of the cosmic-rays which would otherwise swamp the detectors of the 'telescope' with unwanted signals. Another requirement is that the telescope 'views' a large amount of material to give some chance that neutrinos can be detected through their interactions with matter.

In the case of the low-energy solar neutrinos, or neutrinos from a supernova, the aim is to record the neutrinos through their interactions in the detector itself. The telescope is both detector and 'target'. But as neutrino interactions are rare, the target must be large, as we have already seen in the examples of the solar neutrino detectors.

In searching for ultra-high-energy cosmic neutrinos, the emphasis is different. These neutrinos are far rarer; some two million million times more neutrinos from the Sun arrive at each square metre of the Earth's surface than do neutrinos with energies above 10^9 eV. So to produce a useful number of interactions requires a much larger target. Indeed, in this case the aim is to use the Earth itself as the target! Most cosmic neutrinos pass readily through the Earth, but some, especially at higher energies, can react to produce muons. And muons, fortunately, are very penetrating. A muon produced in the ground up to tens of metres beneath a detector can emerge to leave a tell-tale, upward-going track – the signal of the nearby interaction of a neutrino that entered the Earth through the opposite hemisphere.

How do you distinguish upward-going muons from the far more copious downward-travelling ones? One technique depends on measuring precisely the time at which each muon passes through different layers of the detector. A muon travelling downwards will produce signals at the top of the detector slightly sooner than at the bottom of the detector. Conversely, an upward-travelling muon will produce the earliest signals at the bottom of the detector. These differences in timing along a particle's track provide a vital way to reveal its direction. But even in a detector 10 metres tall, the time differences between top and bottom are only about 30 nanoseconds ($\frac{3}{100}$ of a microsecond) for a particle travelling close to the speed of light. So the detector must be capable of cleanly resolving signals spaced closely in time.

These features of underground location, large size, and good timing resolution feature in what may be the first telescope to reveal distant 'point' sources of neutrinos. MACRO, the Monopole, Astrophysics and Cosmic Ray Observatory, is being built in one of the halls of the Gran Sasso Laboratory in Central Italy (see Chapter 6), by a team mainly from Italy and the US, with some members from CERN.

MACRO is much more than a neutrino telescope; it is designed to detect many types of cosmic particle, in particular monopoles, hypothesised

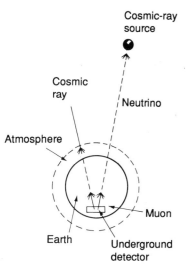

Cosmic-ray source

Cosmic ray

Neutrino

Atmosphere

Muon

Earth

Underground detector

A neutrino observatory can use the Earth itself as a 'target' for the high-energy neutrino interactions; the observatory has then to detect the muon produced, which will appear out of the Earth in virtually the same direction as the neutrino entered it on the other side. Neutrinos detected in this way may come from the interactions of cosmic-rays in the atmosphere on the other side of the Earth, or they may be genuine cosmic neutrinos, produced in similar interactions but in some distance source, perhaps of the kind illustrated on p.199.

particles carrying a single magnetic charge, or pole. It stands 10 metres high, 12 metres wide and when complete will be 72 metres long. The whole observatory is a huge multiple sandwich of layers of detector interspersed with low-radioactivity rock to absorb energy and assist in identifying penetrating charged particles.

Some of the layers are flat containers of liquid scintillator. The scintillator, similar to that used nearly 40 years previously to detect neutrinos for the first time, emits tiny flashes of light when charged particles pass through. Light-sensitive phototubes detect these flashes and provide signals that show to within 1.7 nanoseconds the timing along the track of a particle passing through the detector. This accurate timing means that MACRO should be able to distinguish between the few hundred upward-going muons expected each year and the 10 million or so that will fly downwards through the detectors. The size of the light flashes from the scintillator also gives a measure of the energy of the detected particles.

Other layers of detectors in MACRO pinpoint the tracks of the charged particles, rather than their timing. These are raft-like layers of plastic tubes, each 12 metres long, and 3 centimetres square in section. This type of detector, often named after its inventor E. Iarocci, a member of the team building MACRO, has been used in several experiments that need a relatively simple and inexpensive way of tracking particles over large areas. Each tube has a wire running along its centre and is filled with a special mixture of gases. A positive voltage on the wire sets up an electric field within the tube, so that when a charged particle ionises the gas mixture a discharge occurs. Sheets of metal strips lying across each layer of tubes sense the discharges and produce electrical signals that reveal which tubes have been fired by the particle.

The neutrinos that MACRO detects through upward-going muons will have energies greater than 10^9 eV, and the muons themselves should lie within $1°$ of the direction of the neutrino that produced them. An area of sky $1°$ across, which is about double the size of the Moon, is therefore the smallest region that MACRO can resolve. Calculations show that neutrinos produced in the atmosphere of Earth's southern hemisphere should produce only 1 upward-going muon in MACRO in 10 years for each $1°$ area of sky. This is well below the estimates for the number of neutrinos from specific sources, based on the numbers of gamma-rays that have been detected. Vela X1 and LMC X4, which are X-ray binary systems (two stars, one a pulsar, rotating around each other) that have both been pinpointed as sources of ultra-high-energy gamma-rays, should each give between 5 and 10 muons

The MACRO detector during construction in the Gran Sasso Laboratory in Italy. The detector is built from layers of gas-filled 'limited streamer' tubes, with a layer of liquid scintillator near the top and another near the bottom. It is 12 metres wide and 12 metres tall and in its final form stretches for 72 metres along the experimental hall. (Courtesy G. Giacomelli, MACRO.)

per year in MACRO. These sources in the southern hemisphere lie ideally placed for detection by MACRO which, unlike more conventional telescopes, views the heavens *through* the Earth.

Although MACRO will not be finished in its entirety until 1992 at the earliest, its first complete subsection began operation in the spring of 1989. In a period of three months, 98 000 muons passed through the whole subsection from top to bottom. And one muon travelling upwards was detected, proving at least that MACRO *can* distinguish upward muons from those going down through the detector.

The view 'seen' by the MACRO detector through the cosmic neutrinos it detects, with several potential sources shown. Note that this is in galactic coordinates, so the Milky Way would be spread horizontally across the centre of the oval. Because it detects neutrinos coming *through* the Earth, MACRO does not detect neutrinos from the skies above Italy – this is the portion shaded on the map.

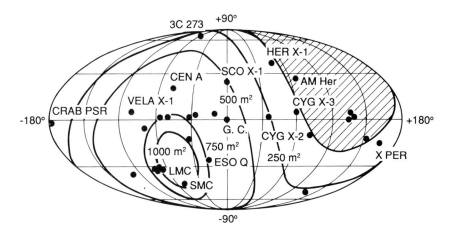

The tracks of a cluster of high-energy muons crossing the first 'supermodule' of MACRO have been reconstructed here. The stars show where each muon has registered in a limited streamer chamber, while the boxes indicate the scintillator traversed by the muons. Notice how one muon has stopped in the detector. (Courtesy G. Giacomelli, MACRO.)

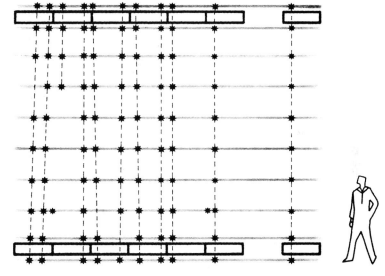

Underwater neutrinos

> We propose setting up apparatus in an underground lake or deep in the ocean in order to separate charged particle directions by Čerenkov radiation[19]
>
> *Moisei Markov, 1960.*

Water is one of the commonest substances on Earth. It abounds in lakes and oceans – and it can be used to detect cosmic neutrinos. High-energy muon-neutrinos travelling through a large mass of water have a small but finite chance of interacting to produce an energetic muon, and because the muon is charged it will emit Čerenkov radiation as it travels through the water. So why not, as Moisei Markov from the Joint Institute for Nuclear Research at Dubna, near Moscow, suggested at the Rochester Conference in 1960, build a detector in a lake or in the ocean, and look for the Čerenkov radiation due to muons produced by cosmic muon-neutrinos? Nature would provide the basic material free of charge, so the detector could cover a large area and have a better chance of capturing one of the rare ultra-high-energy neutrinos.

The ideas of Markov, and others around the same time, later gave birth to the large water detectors, IMB and Kamiokande, which have been described earlier in the chapter. These were built in the early 1980s to detect the possible decay of protons, but they serendipitously became cosmic neutrino detectors on the night of 23 February 1987, when the supernova SN1987A burst into action. Now, however, the original concept of underwater neutrino telescopes is at last beginning to take shape in two new international projects – one (with Markov's encouragement) in the waters of Lake Baikal; the other (the more advanced) in the ocean close to the Hawaiian Islands.

The waters around the Hawaiian Islands are as near perfect as seems possible for a neutrino telescope. The water itself is very clear, clearer even than deep lakes of fresh water. This allows light to travel tens of metres with little absorption, so that detectors to pick up the Čerenkov light can be placed relatively far apart. In addition, there are no strong currents off the islands to disturb the detectors; and their volcanic nature means that the islands have steep sides, which extend underwater, so that the ocean remains deep relatively close to land.

These waters will soon be home to DUMAND, the Deep Underwater Muon and Neutrino Detector, a project that has its roots in discussions at the International Cosmic Ray Conference at Denver in 1973 and later in DUMAND 'workshops' – meetings that brought interested parties together to work through their ideas. The plans, initiated by Fred Reines at the University of California, Irvine, gradually took shape under the later

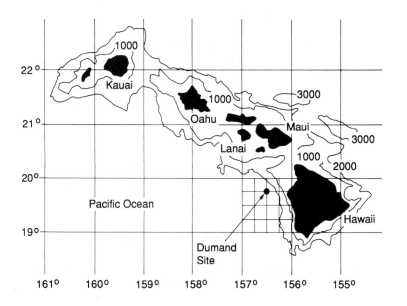

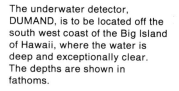

The underwater detector, DUMAND, is to be located off the south west coast of the Big Island of Hawaii, where the water is deep and exceptionally clear. The depths are shown in fathoms.

guidance of John Learned, Vince Peterson, Art Roberts and Vic Stenger, all at the University of Hawaii. The path was by no means easy, as they encountered genuine difficulties in taking particle physics to sea and followed some false turns in the design of the detector. As John Learned recounts:

... we have had many adventures, with high energy physicists learning quite a few painful lessons from the wild and untameable ocean . . . There were formidable potential problems that people had foreseen, such as: monsters of the deep (not many below 2 km), brightly glowing fishes and bacteria (rare, and not much action as long as one does not jostle the instruments about) . . . At first when we talked about DUMAND, our accelerator friends laughed and said we were crazy. (Now they ask why have you not got it operating yet?) . . . It has been a long struggle, but not at all without much fun.[20]

There was also the not inconsiderable task of convincing funding committees that DUMAND was indeed viable, not only technically but scientifically. Various people left the project along the way, while others joined; by 1990 the team included researchers from the Universities of Aachen, Bern, Boston, Hawaii, Kiel, Tohuko, Tokyo, Washington (Seattle) and Wisconsin as well as Vanderbilt University in Nashville and the Scripps Institute of Oceanography.

Selecting a suitable site and investigating the foreseen hazards was only part of the story. At least as much research had to go into the design of a suitable module to detect the Čerenkov light: something that would be robust yet sensitive, large yet manageable, reliable and yet not unduly expensive, and above all, capable of withstanding the high pressures that exist thousands of metres under water. The team designing DUMAND had many calculations to make, many tests to perform, and a number of failures before they eventually completed successful trials at sea with a prototype detector.

In October and November of 1987, members of the DUMAND team went fishing 35 kilometres off the Big Island of Hawaii for a catch of cosmic muons and neutrinos. Their vessel was the *Kaimalino*, a 'SWATH' vessel (small wetted area, twin hull) with a highly stable configuration of twin submerged hulls and a platform supported above the surface. From the well at its centre, a cable that supported various modules plumbed the ocean beneath, extending as deep as 4 kilometres below the surface in some tests. This 'Short Prototype String', as it was called, contained seven optical modules, strung about 5 metres apart, which detected Čerenkov light and transmitted signals to a controller at the end of the string. The controller retransmitted the signals by optical fibre back up to the vessel above for analysis by computer.

A high-energy muon travelling down through the water in the vicinity of the string would emit a cone of Čerenkov light around its path. Some of this light would be picked up by each of the seven modules, the light arriving slightly earlier at the top of the string than at the bottom. The detection of

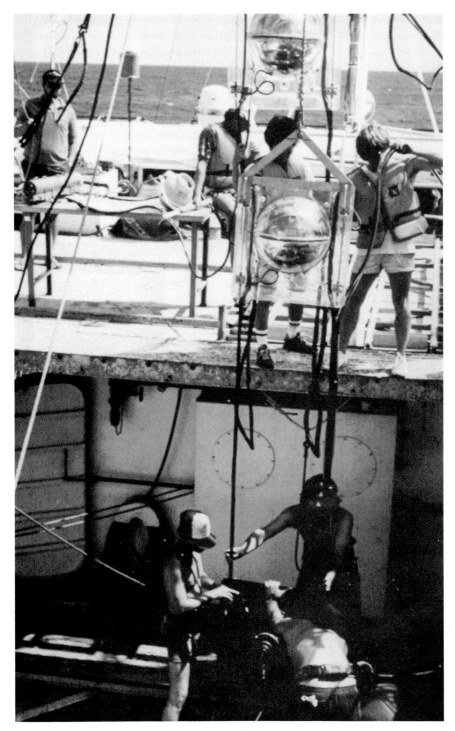

Tests with DUMAND I, a single string of seven optical modules, revealed the potential of the DUMAND project by detecting underwater muons over a surrounding area of 400 square metres. Here members of the team are deploying the string from the vessel *Kaimalino*, and the spherical optical modules are clearly visible. (Courtesy J. G. Learned, DUMAND collaboration.)

DUMAND II will consist of an octagonal array of eight strings of optical modules, with an additional string at the centre. The array will be about 30 kilometres from the coast, in water 4.8 kilometres deep, and a single fibre-optic cable will relay all data collected to the laboratory at Keahole Point on the shore. The modules are placed 10 metres apart on each string, the lowest being only 100 metres above the sea floor.

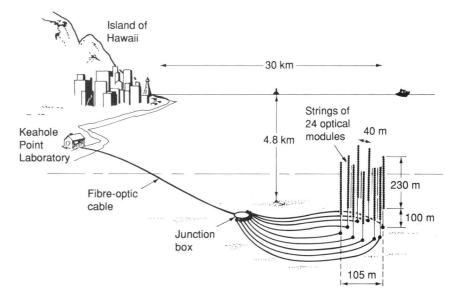

signals with this timing would therefore reveal the muon; the relative intensity of the signals would provide a measure of the distance of the muon's track from the string. In a similar way, a muon travelling up through the water would produce earlier signals at the bottom of the string than at the top. This ability to distinguish between different directions is vital, as Kenneth Greisen had pointed out in 1960 when he discussed the feasibility of a water Čerenkov detector, at a conference at Berkeley, California:

> The Čerenkov detector should be instrumented so as to distinguish the direction of each event. This helps to guard against background, since the horizontal and upwards-directed events are due only to neutrinos.[21]

The tests with the Short Prototype String proved a great success. They not only revealed downward-going muons but even one upward-going muon, possibly due to a neutrino. The results also demonstrated the power of the technique by indicating that the single string of seven optical modules detected muons over an area as great as 400 square metres – comparable to the area of the whole IMB detector.

The success of the Short Prototype String led the team to propose a full-scale, or Phase II, DUMAND in 1988, and this received full approval in April 1990. It will consist of nine strings of optical modules, eight arranged to form an octagon with sides 40 metres long, the ninth string being at the centre of the octagon. Each string will contain 24 optical sensors, spaced at 10-metre intervals, the lowest sensor being only 100 metres above the bottom of the ocean, which at the chosen site is 4800 metres deep. The controller at the base of each string will be connected to a junction box. This will combine the information from the strings, and transmit the data by a fibre optic cable 40 kilometres long to Keahole Point on Hawaii, where there is a research station already used for oceanographic studies. The plans are to begin laying the cable from the site to the shore in the summer of 1992, with the last strings of optical sensors being installed in 1993. As John Learned has remarked,

With funds, diligence, and a bit of luck, the first sources could be reported by 1994.
    . . . It is a dangerous business to make predictions about the detection of inherently unpredictable phenomena, such as the present case in the attempt to begin high energy neutrino astronomy. However, if the gamma observations at [very high energies] are correct, we should expect to see at least a half dozen neutrino point sources with DUMAND II.[22]

Only the most energetic downward-going muons will reach the depths at which the sensors will lie, to produce signals at an estimated rate of 3 per minute. The numbers due to atmospheric neutrinos will be far smaller, in the region of about 10 a day over all angles. Any distinct 'point' source of high-energy neutrinos in the heavens should appear clearly against this background of atmospheric neutrinos, provided that during the course of a year it emits enough high-energy neutrinos within an angle of 1° for ten or more to interact in the water above DUMAND.

Meanwhile, at the deepest freshwater lake on Earth, Lake Baikal in Eastern Siberia, another team of researchers has been making progress with NT-200, the proposal for a 'neutrino telescope with 200 photomultipliers'. From 1984–5 the international team, which includes physicists from Irkutsk, Moscow and Tanzsk, as well as Budapest and Berlin, has successfully tested strings of up to nine optical modules at a depth from 850–1350 metres. In this case, each module consisted of four relatively small 15-centimetre diameter phototubes, two looking up and two looking down. The team found that they could deploy the strings relatively easily in winter, as they had a natural platform provided in the form of the thick layer of ice that covers Lake Baikal from February to April. And they could lay the cables to carry signals to the shore by feeding them through a slot in the ice cut by a giant saw towed by a sledge!

In the final design, the team plans to use eight strings of six optical modules (four phototubes per module) which will together cover an area of 3000 square metres. The phototubes will be larger than in the tests with hemispherical heads 35 centimetres in diameter. The aim is to install these by 1993, and then to begin the search for upward-going muons, the tell-tale sign of high-energy neutrinos from the other side of the Earth.

Twenty-five years ago, after they had first detected atmospheric neutrinos, Fred Reines and J.P.F. Sellschop wrote:

It is difficult to predict the future of neutrino astronomy, but it seems likely that what began 35 years ago as an ingenious apology for an untraceable loss of energy will eventually enable physicists not only to probe the fundamental forces of nature on the subatomic scale but also to reveal much about processes at the center of the sun and perhaps more distant celestial events as well.[23]

On Lake Baikal, members of the NT-200 team take advantage of the thick layer of ice during the winter months to deploy their test string for an underwater neutrino detector. Here a pair of optical modules, each containing a phototube 35 centimetres diamater, is being lowered into the water below the ice. The spheres are made from glass but are designed to withstand the pressure at depths of up to several kilometres. (C. Spiering, Institut für Hochenergiephysik, Zeuthen.)

It is now 60 years since Pauli apologetically introduced the neutrino, and the 'distant celestial event' of SN1987A has at last shown the potential of neutrino astronomy. The future, as in any field of pure research, remains difficult to predict, but to quote Reines and LeRoy Price writing more recently:

Our field of neutrino astrophysics has made a notable, serendipitous beginning. *Now is the time to be bold and move strongly forward.*[24]

In the beginning . . .

> The effort to understand the universe is one of the very few things that lifts human life a little above the level of farce, and gives it some of the grace of tragedy.[25]
>
> *Steven Weinberg, 1977.*

Neutrinos stream through space, produced in nuclear fusion in stars, in the cataclysmic deaths of stars, in exotic cosmic accelerators. There is also an older source, long since evolved from its early state, but its neutrinos should remain, relics of a bygone age. This source is the primordial Universe, a mere second old. All neutrinos that existed then should still permeate the Universe, a few hundred in every cubic centimetre of space, witnesses to the Universe's evolution.

The present 'Standard Model' of the Universe is known as the 'Hot Big Bang'. According to this theory, the Universe is expanding from a state of possibly infinitely high density and temperature, in which it existed at time zero, the instant of the Big Bang. As the Universe expanded, it cooled, and while we do not yet have the correct theories to describe the initial state of the Universe, we can use the physics of environments that we do understand to describe what the Universe was like at different times as it cooled.

During the time since the Big Bang, some 15 billion years ago, matter in the Universe has evolved to the matter we presently observe in the atoms and nuclei of our planet, of the Sun, and of our galaxy and many others. More surprisingly, perhaps, it appears that the forces that act upon this matter, and govern its behaviour, have also evolved as the Universe has cooled. In the present Universe, the strong force in an atomic nucleus is strong enough to overcome the electrical repulsion between protons; and the weak force is weak enough for the Sun to have kept shining for several billion years. In the high-temperature environment of the very early Universe, however, the forces had more equal strengths, and weak interactions, involving W and Z^0 particles, would have been as frequent as strong interactions, involving gluons. It therefore seems likely that at the time of the Big Bang, there was a single 'unified' force, including not only strong, electromagnetic and weak interactions but also gravity. Then, as the Universe cooled, the fundamental forces, which now appear so different, separated out from this unified force, rather as fat separates out from a more jelly-like substance when the homogeneous 'dripping' from roast meat cools down.

In its first moments, immediately after the Big Bang, the Universe would have consisted of an exceedingly high-temperature 'soup' of elementary particles, containing quarks and leptons, together with the force-carrying photons, gluons, W and Z particles, and also, possibly, other particles that are still to be discovered. At these very high temperatures, the average energy of all the particles would have been too high to form stable protons or neutrons, let alone nuclei or atoms. Only as the Universe cooled, with the average energy falling with the temperature, would at first protons and neutrons have formed, and later nuclei, followed by atoms. Eventually, the atoms would have clumped together to form stars and galaxies, to make the Universe we see today.

The expanding Universe

> All we have to do is to accept relativistic formulae for the universal expansion, and empirical data concerning various nuclear reactions, and see if the calculations lead to a result which resembles the observed abundances of known atomic species.[26]
>
> *George Gamow, 1952.*

The theory of the Hot Big Bang originated in the 1940s, in an attempt by George Gamow to invent what he referred to as a 'primordial pressure cooker' in which basic ingredients were cooked to form elements in the proportions observed in the present Universe. Gamow was a Russian physicist who emigrated to the US, where he became professor at the George Washington University in Washington, DC. In the late 1920s, he proposed the mechanism of quantum tunnelling, which in 1935 he used to explain how thermonuclear reactions could fuel stars like the Sun, as we saw in Chapter 6. And in the 1940s, after working on theories of how elements are formed in stars, he turned to a problem that was challenging a number of people at the time – the origin of elements in the primordial Universe.

Gamow's original paper on the formation of elements in the early Universe appeared in 1946, and this was followed in 1948 by a famous paper written with his student Ralph Alpher, on 'The origin of the chemical elements'. A typically playful move by Gamow was to add, with permission, the name of Hans Bethe (*in absentia*), so that the authors could be listed as Alpher, Bethe and Gamow. The words *in absentia* somehow became lost, however when the paper was published in the 1 April issue (purely by chance) of *The Physical Review*, volume 73. (Much to Gamow's chagrin, Robert Herman, who worked with Gamow and Alpher on later papers, refused to change his name to Delter!)

The aim of Gamow and his colleagues was to explain the abundances of elements observed today, with one necessary condition: that the Universe is expanding. Edwin Hubble, an American astronomer, had produced the first evidence for an expanding Universe in 1929. Observations made from around 1912 onwards revealed that a number of galaxies are moving away from us. Hubble's step forward was to assign distances to those galaxies and to show that the farther a galaxy is from us, the more rapidly it is moving away. This simple proportionality became enshrined in what is now known as Hubble's law.

The velocity–distance relationship takes on a great significance for the Universe as a whole if we assume that we are not privileged to live in a particularly special part of the Universe; in other words, if we infer that *all* galaxies are rushing away from each other, in accordance with Hubble's law.

The proportionality between velocity and distance is then precisely what we would expect if the Universe is expanding. In his book *The Expanding Universe*, Arthur Eddington described the conclusions that he and his contemporaries drew immediately from Hubble's discovery:

... if you ask what is the picture of the universe now in the minds of those who have been engaged in practical exploration of its large-scale features ... their picture is the picture of an *expanding universe*. The super-system of the galaxies is dispersing as a puff of smoke disperses.[27]

One consequence of interpreting Hubble's law in terms of an expanding universe, is that it provides an 'age' for the Universe. If all galaxies are moving away from each other, in the past they must all have been closer together, and at some time, which we can define as the beginning of the Universe, they must have all been in the same place. From the measured value of Hubble's constant, we can calculate how long ago the galaxies were all together, and find an answer of between 10 and 20 billion years. This is happily consistent with estimates for the age of the Earth, for example, which are based on entirely different measurements of radioactivity within rocks. (This was not always so, however, for the early values for Hubble's constant implied a Universe younger than the Earth! Later, improved astronomical observations led to the currently accepted value and a self-consistent theory – the Big Bang.)

The expansion of the Universe is not in itself evidence for the Big Bang; rather it is a basic axiom of the theory. For evidence we must look elsewhere. Clues as to where to look were embedded in a paper published in the journal *Nature* in 1948 by Gamow's colleagues, Alpher and Herman. This paper improved upon an attempt by Gamow, published earlier in the same volume of *Nature*, to follow the evolution of matter and radiation in the Universe from a time of about 1 second through to the formation of galaxies at a time when the densities of matter and radiation were about equal. The starting point was a very dense and therefore high-temperature Universe made entirely of neutrons basking in a bath of radiation. The neutrons would decay and so create protons, and the protons and neutrons would then form deuterons. The approach was then to calculate the densities of matter and radiation at 1 second and extrapolate to the time when they were equal.

Gamow's treatment not only contained a few errors, but his extrapolation led to too long a timescale for atoms to form – indeed, the result was greater than the accepted age of the Universe! Alpher and Herman found that they could follow the densities through without extrapolation, and they obtained a more reasonable time scale, suggesting that radiation ceased to dominate matter after about 10 million years. In addition, they estimated the present temperature of radiation in the Universe to be about 5 degrees above absolute zero:

The temperature of the gas at the time of condensation was 600 K, and the temperature in the Universe at the present time is found to be about 5 K[28]

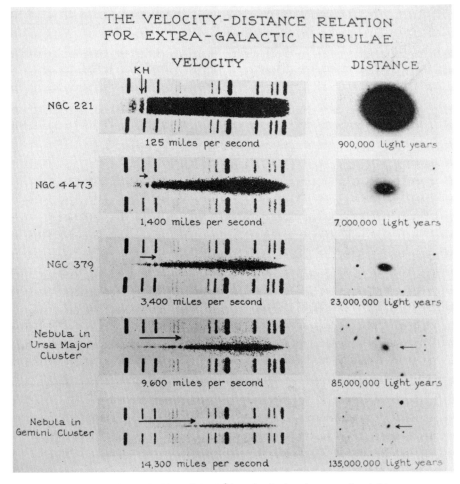

THE VELOCITY-DISTANCE RELATION
FOR EXTRA-GALACTIC NEBULAE

	VELOCITY	DISTANCE
NGC 221	125 miles per second	900,000 light years
NGC 4473	1,400 miles per second	7,000,000 light years
NGC 379	3,400 miles per second	23,000,000 light years
Nebula in Ursa Major Cluster	9,600 miles per second	85,000,000 light years
Nebula in Gemini Cluster	14,300 miles per second	135,000,000 light years

The increasing recession velocities of the series of galaxies shown on the right are revealed by a careful study of their spectra, shown on the left. The observed spectrum of light from each galaxy is the elongated dark shape; the dark lines above and below are spectral lines from a standard laboratory helium source. In each galactic spectrum a pair of bright lines (H and K), due to absorption by calcium are visible, but they occur more and more towards the red end (towards the right) of the spectrum, due to the increasing speed at which the galaxy being observed is receding from us. This recession velocity is directly related to the distance of the galaxy, providing strong evidence that the Universe is expanding. (*The Realm of the Nebulae* by Edwin Hubble (OUP, London, 1936).)

In the process of expansion, the high-energy gamma-ray photons of the hot early Universe would have given way by the time of the present-day Universe to low-energy 5 K radiation in the form of microwaves.

This prediction went more or less unrecognised, probably, as Steven Weinberg has discussed in his book *The First Three Minutes*, mainly because at the time it was difficult to take seriously *any* theory of the very early Universe. As Weinberg says:

... the first three minutes are so remote from us in time, the conditions of temperature and density are so unfamiliar, that we feel uncomfortable in applying our ordinary theories of statistical mechanics and nuclear physics. ...

Gamow, Alpher and Herman deserve tremendous credit above all for being willing to take the early universe seriously ... Yet even they did not take the final step, to convince the radio astronomers that they ought to look for a microwave radiation background.[29]   1933

c '53   It was two decades later that two physicists, working for the Bell Telephone Laboratories in New Jersey, discovered microwave radiation, corresponding to a temperature of about 3 K, which appears to permeate the Universe. Their paper, modestly titled 'A measurement of excess antenna temperature at 4080 Mc/s', appeared in *The Astrophysical Journal* in 1965. This was the first piece of solid evidence for the Hot Big Bang. Arno Penzias and Robert Wilson had discovered the after-glow of the initial high-temperature state, cooled to a mere 3 degrees above absolute zero during the continuous expansion of the Universe over 15 or so billion years. And, says Weinberg:

The most important thing accomplished by the ultimate discovery of the 3 K radiation background in 1965 was to force us all to take seriously the idea that there *was* an early universe.[30]

However, many astrophysicists argue that some of the best evidence for the Hot Big Bang comes from following Gamow's original intentions – that is, through using a model of the formation of elements in the early Universe, to predict the relative amounts of lightweight elements that we observe in the Universe today. It is remarkable indeed that these studies of primordial nucleosynthesis – the process of building light nuclei from protons and neutrons – can correctly yield the abundances of elements observed some 15 billion years later, the more so as these abundances differ by as much as a billion from one element to another.

The cosmic cooker

The number of elementary particles must be limited, otherwise the Universe would be different from the one we know.[31]
                              David Schramm and Gary Steigman, 1988.

Big-Bang nucleosynthesis can explain what happened in the Universe after about the first hundred thousandth of a second. By this time, the Universe had expanded and cooled to around 1000 billion degrees, and protons and neutrons would exist, 'condensed out' from an earlier high-energy broth containing quarks and gluons. There would also be neutrinos, electrons, muons, and their antiparticles, as well as photons, all dominating the protons and neutrons by a factor of around a billion to one. At first, the neutrons and protons would be in equilibrium; protons would be able to change to neutrons through weak interactions with antineutrinos as readily as neutrons could turn to protons through similar interactions with neutrinos. But as the Universe continued to expand and cool, the weak interactions

Arno Penzias (left) and Robert Wilson together with the horn antenna with which they discovered the cosmic microwave background radiation – the present-day relic of the Hot Big Bang. (Bell Laboratories.)

slowed down and the transformation of the neutrons to protons through normal beta-decay would begin to dominate. The number of protons would therefore begin to exceed the number of neutrons.

By the time the Universe was about 1 second old, its temperature had fallen to 10 billion degrees and the average energy would have fallen to about 1 MeV. At this point, the rate at which collisions through weak interactions occurred would be less than the rate of expansion of the Universe. From this time on, the only weak interaction left for the protons and neutrons was the beta-decay of the neutrons. And already there would be less than one neutron for every three protons. Had nothing else happened as the Universe continued its inexorable expansion, the remaining neutrons would simply have decayed away during the next 15 minutes. But as it happens within nearly 2 minutes or so the neutrons began to become locked within simple atomic nuclei; nucleosynthesis, the creation of the elements, had begun, 100 seconds after the initial Big Bang.

The temperature of the Universe had now fallen to a billion degrees, which was low enough for a proton and a neutron to come together to form a nucleus of deuterium (heavy hydrogen). Before this, at higher temperatures, the radiation permeating the Universe would have been energetic enough to split deuterium nuclei as soon as they formed; now they could survive long enough to participate in further reactions. The deuterium nuclei could interact with an additional proton or neutron to form nuclei of helium-3 or of tritium (hydrogen with two neutrons). The helium-3 and the tritium could react, in turn, with a proton or neutron, respectively, to form the very stable nucleus of helium-4 (two protons, two neutrons). Lithium-7 was also produced, in small amounts, in reactions between helium-4 and tritium. It was crucial at this stage that the temperature, although low enough to make deuterium, was at the same time still high enough to allow the positively charged nuclear fragments to come together, overcoming their mutual electrical repulsion, through Gamow's quantum tunnelling.

No elements heavier than lithium could be made in this primordial melting-pot, for no stable nuclei are made in interactions between helium-4 and protons or neutrons, or indeed between two helium-4 nuclei. Instead, it takes the high temperatures and densities that exist at the centres of stars like the Sun to bring three helium-4 nuclei together to make carbon-12. It was to be more than 300 000 years before the first stars could ignite, and additional new elements be born.

The spate of element building in the early Universe lasted a little more than 3 minutes. By this time all the neutrons that had existed after the first second were locked mainly inside nuclei of helium-4, with some smaller amounts in deuterium and helium-3, and still smaller amounts in lithium-7. The relative abundances of the lightest nuclei, hydrogen, deuterium, helium

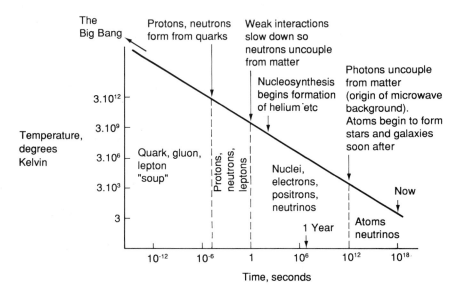

Since the initial Hot Big Bang, the Universe has expanded and cooled to the 3 K it is today. As it cooled, it passed through various stages – helium and other light nuclei forming once the ambient energy was sufficiently low for the nuclei to remain bound together, when the Universe was less than three minutes old. But it was 300 000 years before the Universe was cool enough for electrons to bind with nuclei, to form atoms.

and lithium, from which the present Universe would evolve, were now fixed. However, it would be another 300 000 years before atoms could form, when the Universe would have cooled to some 3000 degrees and the nuclei could at last capture electrons and keep them in orbit. The stars and galaxies would have begun to form soon after.

The electrons, like the quarks and gluons inside the nuclear particles, would have existed since the earliest instants of the Universe. When the neutrons and protons 'condensed out', after the first hundredth of a second or so, the primordial cauldron contained not only quarks and gluons, but also electrons, neutrinos, and, of course, photons. At this stage the electrons and neutrinos were in equilibrium with the photons. In other words, photons could create electron–positron pairs as readily as they annihilated back to photons; similarly neutrinos and antineutrinos were kept in equilibrium through their weak interactions.

In 1969, V.F. Shvartsman, at the Moscow State University, pointed out that the amount of primordial helium that exists today – that is, helium made in the early Universe as opposed to helium made in stars, as described in Chapter 6 – depends on the density of lightweight particles during this period prior to nucleosynthesis. In particular, Shvartsman referred to 'DZPs', or 'difficult-to-observe particles with zero rest mass',[32] which included 'still unknown ultra-weakly-interacting particles left from the superdense phase'. In a paper published in *JETP Letters*, he showed that the present density of such particles had to be less than five times the density of photons in the microwave background, if nucleosynthesis was going to yield a helium abundance of less than 40 per cent, the figure that was then accepted.

The reason that such a limitation arises is due to the crucial interplay between the rate at which the early Universe expanded and the rates of the fundamental interactions. The more types of particle that were in equilibrium with the radiation at a given time, the greater the energy density at that time; and the greater the energy density, the higher the rate of expansion. If the Universe expanded faster just before nucleosynthesis set in, then there would be more neutrons in proportion to protons at the point at which the expansion became more rapid than the weak interactions between protons and neutrons. Moreover, there would have been less time for the neutrons to decay before the Universe became cool enough for deuterium nuclei to form. The result of a more rapidly expanding Universe would have been more neutrons at the time of nucleosynthesis and therefore, ultimately, more helium-4.

The work of Shvartsman, pursued later by many others, showed that particle physicists are not at complete liberty to invent particles to explain any unexpected effects they discover. Not only must any new particle fit into a theory of particle physics, it must also behave in such a way that cosmologists can continue to explain the development of the Universe from its initial Big Bang to the present observed state. In particular, as Gary Steigman, David Schramm and James Gunn pointed out early in 1977, the number of types of neutrino that existed before nucleosynthesis began is intimately related to the amount of primordial helium-4 that we measure today.

These three American astrophysicists were motivated by the announcement in 1976 of the discovery of a third type of charged lepton – the tau – by Martin Perl and his colleagues (see Chapter 4). Steigman, Schramm and Gunn realised that the existence of a third type of neutrino, the uncharged partner for the tau, would have speeded up the expansion of the early Universe, and hence led to the presence now of more helium-4. In a paper submitted to *Physics Letters* at the end of November 1976, and published in January 1977, they showed that there could be at most seven types of lightweight neutrino (including electron- and muon-neutrinos) if their calculations of nucleosynthesis in the Big Bang were to reproduce estimates of the relative proportion of primordial helium-4 in the Universe as a whole. However, they concluded with the following caveat:

Of course, it should be remembered that the standard big bang model is by no means rigorously verified but is merely the simplest model which seems consistent with observations.[33]

In the following years, as more data accumulated and measurements became more refined, the Big Bang theory continued to be the only astrophysical model capable of producing the correct amounts of primordial helium-4 and deuterium, and its claim to legitimacy strengthened. In June 1984, Schramm and Steigman, together with Jonman Yang, Michael Turner and Keith Olive, were able to begin a major article in *The Astrophysical Journal*, with the words:

The almost universal acceptance of the standard (i.e. simplest) hot big-bang model . . . rests, in large part, on the success of this model in accounting for the abundances of the light elements, particularly helium-4 and deuterium.[34]

The article showed, among other things, that the number of lightweight neutrinos must be limited to four at most, the fourth type being barely allowed. In this way, nucleosynthesis for some time provided a far more stringent limit on the possible number of neutrino types than experiments in particle physics. Only in 1989 did the experiments on the LEP accelerator at CERN, and on the SLC machine at SLAC, yield a tighter limit, when measurements of the decays of Z particles produced in electron–positron collisions revealed that there can be no more than three types of lightweight neutrino.

Relic neutrinos

The most dramatic possible confirmation of the standard model of the early universe would be the detection of this neutrino background[35]

Steven Weinberg, 1977.

The very early Universe was 'radiation dominated'; in other words, the main contributions to its energy came from photons and other particles moving at almost the speed of light. By contrast the present Universe is 'matter dominated'. The energy of the photons has fallen as the Universe has expanded and cooled, so that now most of the Universe's energy is in the form of low-energy but massive particles, the protons and neutrons of nuclear matter. During the early, radiation dominated phase, the Universe behaved like a 'gas' of radiation in thermal equilibrium. This makes it relatively straightforward to calculate the density of photons and other particles that existed at a given temperature and therefore at a particular time. The density depends in part on the spin of the particles, photons having a spin of 1 unit, neutrinos a spin of $\frac{1}{2}$. This implies that there would have been $\frac{3}{2}$ as many neutrinos of a single type as there were photons during the first second or so of the Universe, if the neutrinos are Dirac particles, with distinct particle and antiparticle, as discussed in Chapter 4. (If they are Majorana particles, with particle and antiparticle the same state, there would be half the number of neutrinos, that is $\frac{3}{4}$ as many neutrinos as photons.)

When the Universe had cooled to some 10 billion degrees, after about one second, the rate of weak interactions between neutrinos became slower than the rate at which the Universe was expanding. At this point the neutrinos were no longer in equilibrium with the other particles, and from then on the distribution of neutrinos simply thinned out as the Universe expanded. Five hundred thousand years later, a similar 'decoupling' happened to the photons in the Universe. By this time the Universe had cooled to about 3000 degrees, corresponding to a photon energy of 0.3 eV. The photons then had insufficient energy to pull electrons apart from the atomic nuclei that had formed during nucleosynthesis, and neutral atoms, mainly of hydrogen and helium, began to form. Since then the distribution of photons left over from the early Universe has expanded with the Universe, cooling to a temperature of 2.7 K. These are the relic photons that Penzias and Wilson detected as the microwave background radiation, and measurements show that there are some 400 per cubic centimetre of the present Universe.

It might seem that the number of relic neutrinos of a particular type should be simply $\frac{3}{2}$ (or $\frac{3}{4}$) the present number of relic photons. However, back in the early days of the Universe something changed the number of photons soon after the neutrinos went out of equilibrium at a temperature around 10 billion degrees. At temperatures below around 3 billion degrees the photons no longer had sufficient energy to create pairs of electrons and positrons, so these particles ceased to be in thermal equilibrium. However, annihilations between electrons and positrons could continue. This increased the number of photons while decreasing the number of electrons and positrons, and in fact the number of photons should have nearly trebled. This in turn implies that the number of neutrinos of a given type should now be only a half (or a quarter) the number of photons.

This is still a large number of neutrinos. Given that there appear to be three types of neutrino, there should be between 300 and 600 relic neutrinos in every cubic centimetre of space. This, of course, assumes that the neutrinos are stable, or at least have lifetimes greater than 15 billion years, the age of the Universe. Whether we can ever detect this 'neutrino background' will depend on the ingenuity of physicists. But what if some, or all, of these neutrinos are indeed relatively stable, and have mass.

Will the expansion of the Universe ever come to a halt? The answer depends on the amount of matter that there is in the Universe, or more specifically on the energy density. If the density is great enough, the expansion will eventually cease and the Universe will begin to contract; the Universe will be 'closed'. Alternatively, the Universe could expand forever if the density is not enough to halt the expansion; this would be an 'open' universe. A third possibility is that the density is just sufficient to bring the expansion to a stop, after an infinite amount of time, but not quite enough to set off contraction. In this case, the Universe would be 'flat'.

The density required for a flat Universe is called the 'critical density' and its value can be calculated from a knowledge of Newton's gravitational constant and of Hubble's constant, which relates the velocity at which a galaxy recedes to its distance. Hubble's constant is known only to within a factor of two, but this is sufficient to give a value for the critical density of between about 3 and 10 keV per cubic centimetre. If there are 400 relic photons per cubic centimetre with a present energy of 0.0003 eV then their contribution to the overall energy density of the Universe amounts to 0.12 eV per cubic centimetre, a factor of up to 100 000 below the critical density.

If neutrinos have no mass at all, then the contribution of relic neutrinos to the overall density will be equally paultry. But a small mass could make all the difference. With a mass of 30 eV, say, 100 neutrinos per cubic centimetre would result in an energy density of 3 keV per cubic centimetre, close to the range of values estimated for the critical density. The amazing possibility arises that neutrinos, pieces almost of nothing, could determine the fate of the Universe; they could in a sense be masters of the Universe.

But the masses of neutrinos are infamously difficult to measure. In 1981, as Chapter 3 describes, results from an experiment by a team of physicists in Moscow suggested that the mass of the electron-neutrino might lie somewhere in the region of 20–40 eV. This triggered great excitement among cosmologists and particle physicists alike. However, by 1991, ten years later, other experiments had shown that the mass of the electron-neutrino must be less than 15 eV, and that there is probably something wrong in the analysis of the measurements made in Moscow. At present, laboratory experiments set much higher limits on the masses of the muon- and tau-neutrinos, of 250 keV and 35 MeV respectively.

Cosmological arguments, in fact, provide tighter limits on the possible masses of the neutrinos. Astronomical measurements of the rate of expansion of the Universe, itself a difficult quantity to measure, suggest that the density cannot be more than double the critical density. In this case, the total mass of the three neutrino types must not be more than around 100 eV.

But what if neutrinos do not have long lifetimes, of 15 billion years or more? We know, for example, that electron-neutrinos must have a lifetime of at least 500 000 seconds (about 6 days), otherwise neutrinos from the supernova SN1987A would not have been detected here on Earth. However, that is a very short time compared with the time since the Big Bang; and what of the other types of neutrino? If neutrinos do decay, the problem changes to one of considering the effects of whatever it is that the neutrinos decay into; conservation of energy implies that they cannot simply disappear.

A number of observations make it unlikely that neutrinos could decay to photons. Relic neutrinos that decay to photons after the time when atoms began to form, and the Universe became transparent to photons, would have produced radiation over and above the microwave background radiation; such radiation is not observed. Neutrinos that decayed to photons before the formation of atoms would have influenced the shape of the spectrum of background radiation; such an effect is not observed. Neutrinos from supernovae decaying into photons would produce a background of high-energy photons; such a background is not observed.

These observations, or nonobservations as the case may be, impose severe constraints on the combinations of lifetimes and masses that are possible for neutrinos decaying into photons. It could be that neutrinos decay in some other way, but attempts to make theories consistent with the various observations almost always lead to effects that lie beyond the Standard Model of particle physics. In addition, arguments based on the energy density of the Universe, which ensure that the present density is no greater than is consistent with observation, show that certain combinations of masses and lifetimes are completely excluded, even if neutrinos decay in some way that theorists have so far failed to imagine.

Neutrinos as dark matter

> From all the many dark matter candidates proposed in recent years, neutrinos are the only one that definitely exist, and even so, we do not know if they are massive, as they should be.[36]
>
> *Graciela Gelmini, 1988.*

One problem in cosmology that is intimately related to the question of the energy density of relic neutrinos is the mystery of the missing or 'dark matter'. This dark matter enters cosmological discussions at two different levels. First, astrophysicists know from measurements of the dynamics of spiral galaxies that the luminous matter they observe in a galaxy must form only about 10 per cent of the total galactic mass. To account for the orbital velocities of stars and gas in one of these galaxies, the galaxy must be surrounded by a spherical region, or halo, of nonluminous matter, extending at least ten times further than the visible disc of the galaxy.

The energy density of the luminous matter itself is about 1 per cent of the critical density. Taking the galactic dark matter into account brings the energy density up to 10 per cent of the critical density. Similar dynamical arguments about the motion of 'superclusters' of galaxies introduce additional dark matter and lead to an estimate of the overall density of the Universe of 20 per cent the critical density.

Surprisingly, perhaps, this value is close enough to 100 per cent to cause cosmologists problems, for the following reason. In an expanding Universe, the density must have been higher in the past, although clearly it cannot have been much greater than the critical density. However, if we extrapolate back in time working from a present density that is 20 per cent of the critical density, we find that the original density must have been very close to the critical density; indeed, so close as to be within 50 decimal places of the critical density!

It seems remarkable that the Universe should have originated with what cosmologists call such 'fine tuning'. An arguably preferable option is to assume that the Universe began with, and still has, the exact critical density.

Measurements of motion in clusters of galaxies, such as the Virgo cluster shown here, provide evidence for the existence of 'dark matter' – invisible matter, which can be detected in no way other than through its gravitational influence. (Royal Observatory Edinburgh.)

Moreover, a currently popular modified version of the Big Bang model, called the 'Inflationary Universe', specifies that the Universe must indeed be of the critical density. The model of the Inflationary Universe has great appeal among cosmologists because it solves a number of problems that otherwise arise in the very early Universe. I shall not begin to try to explain the model in detail, but suffice it to say that it has provided a great impetus to the belief that the Universe has the exact critical density. But if that is so, we are left with the problem that we see directly only 1 per cent of the matter in the Universe. The remainder, 99 per cent of the Universe, is invisible, 'dark matter'. What can it be?

Many possibilities have been discussed, and many new particles hypothesised, and it may be that the answer lies in some combination of different forms of invisible matter. One choice, which has fluctuated in popularity over the past decade, is that neutrinos form the dark matter.

As explained above, stable neutrinos with a mass around 30 eV, remnants of the early Universe, could provide the critical density. But this solution to the problem of the dark matter went out of favour, when it emerged that as the Universe evolved matter would first cluster together on a very large scale, much larger than the scale of galaxies, and that galaxies would form later, in fact too late. But during the 1980s, astronomers found increasing evidence for the existence of huge structures in the Universe, on a scale corresponding to vast numbers of galaxies. Moreover, cosmologists also discovered a means to speed up the production of galaxies in a Universe dominated by lightweight neutrinos. Once again these neutrinos have become a possible candidate for the dark matter.

It is also conceivable that the dark matter could be heavier neutrinos – for example tau-neutrinos – with a mass around 1 keV, provided that these neutrinos are unstable. But in this case the problems connected with forming galaxies as the Universe evolves seem more intractable, and no one has thought of a mechanism that will allow the present Universe to be built. Another possibility is that there are neutrino-like particles, yet to be discovered, that are at least as heavy as protons. The existence of new types of heavy neutrino is not ruled out entirely by the production of helium-4 in the first three minutes, since this constraint refers to neutrinos that are lightweight, less than about 25 MeV in mass. Moreover, with heavy neutrinos as dark matter there is no problem with the formation of galaxies in the Universe, as the heavy neutrinos tend to make matter clump together first on a galactic scale. Only later would the galaxies come together in clusters and superclusters. However, such neutrinos should be produced in the decays of the Z^0 particle, and so influence its lifetime in a measurable way; the measurements on Z particles at CERN (see Chapter 4) appear to rule out this possibility.

The nature of the dark matter in the Universe, which undoubtedly exists at the very least at the level necessary to explain the dynamics of galaxies, is one of the most important scientific puzzles of the late twentieth century. Its significance extends well beyond the bounds of observational astronomy, to

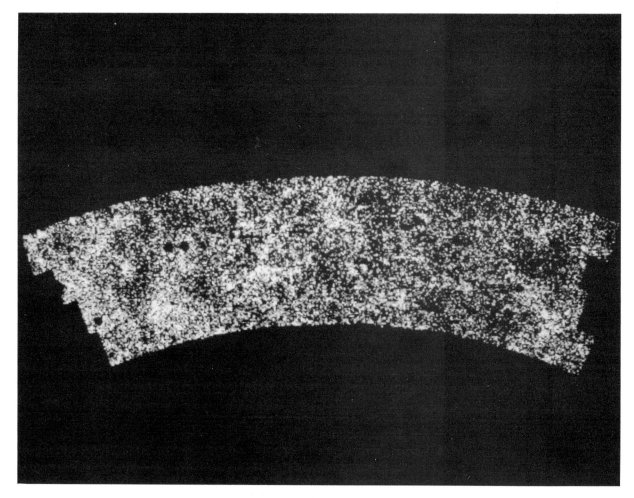

This mosaic shows the distribution of 80 000 faint galaxies, in the general direction of the South Galactic Pole, as measured by the UK Schmidt Telescope, and classified in the Edinburgh–Durham Galaxy Catalogue. The galaxies are clearly not evenly distributed across the sky, but are clustered together in places, leaving large, emptier patches in other regions. Such irregularity is difficult to reconcile with the uniformity of the microwave background across the Universe, but it becomes easier to explain in theories which assume that the Universe contains large amounts of invisible, dark matter. (Royal Observatory, Edinburgh.)

cosmology and to particle physics. The problem epitomises the way in which studies that at one time seemed to be separately related to the vastly different scales of the atom and of the Universe, are, in fact, intimately intertwined. Discovering the nature of the dark matter could be one of the first major scientific achievements of the twenty-first century. And then we shall see whether Pauli's humble neutrino has yet another unseen role in moulding the world, and the Universe, in which we live.

8

Moonbase neutrino

Let me describe a dream of mine for the future . . .
Masa-Toshi Koshiba, 1987.

IT IS NOW MORE THAN 60 years since Pauli hesitatingly produced his 'desperate remedy' for the problems with nuclear beta-decay; and 40 years since Fred Reines and Clyde Cowan began their pioneering attempts to detect neutrinos. It is 30 years since the muon-neutrino was proved to exist in the first experiment with a neutrino beam at a particle accelerator; and 20 years since the first solar neutrinos were detected. In an age when many scientific advances occur at a breathtaking pace, the very elusiveness of neutrinos has made progress in learning about them seem almost ponderously slow. Progress with neutrinos, at least in experiments, has required an enormous amount of patience and care. And imagination.

Throughout this book we have seen how experiments with neutrinos have often pushed techniques to the very limits, in one of the best traditions of pure scientific research. But while the rewards may have been slow in coming, they have been tremendous. The first evidence that the electromagnetic and weak forces are different facets of one underlying force; confirmation of the existence of quarks and gluons within the proton and the neutron; clear support for theories of the death of a star as a Type II supernova – these achievements have all resulted from a dogged determination to persevere with difficult experiments, and an ability to dream of making what seems impossible move into the realms of possibility.

In these closing pages, I would like to convey some flavour of the dreams that neutrinos continue to inspire in physicists – dreams that may often seem impractical today, but which some time in the twenty-first century may become reality.

I have quoted Masa-Toshi Koshiba above, writing in an article on

'Observational neutrino astrophysics' in 1987, in the wake of the spectacular observation of the supernova SN1987A. Koshiba's experiment, Kamiokande II, not only detected neutrinos from the supernova, but was also the first to detect solar neutrinos in 'real time', which could be shown to point back to the Sun. In the spring of 1991, the team received approval for Super Kamiokande, a larger version of their water Čerenkov detector, which should be able to detect sufficient solar neutrinos to monitor the temperature of the Sun's core to 1 per cent accuracy over a week.

However, in his article in 1987, Koshiba was already looking still further ahead, to a world-wide network of similar 'super' neutrino observatories, all with good timing accuracy. According to Koshiba, these would together reveal *daily* variations in the Sun's core temperature at the level of 1 per cent. In addition, the detectors would provide a 'supernova early warning system', as they could detect bursts of neutrinos from supernova explosions well in advance of the first optical signals. Koshiba also envisioned a similar network of high-energy cosmic neutrino detectors installed in lakes across the world, to monitor the entire sky continuously.

Inspired by a somewhat different dream, Francis Halzen and colleagues in the United States have begun to investigate a promising method for extending studies of the highest energy cosmic neutrinos. They envisage a detector that couples the cheapness of the medium used in an underwater detector with the mechanical stability of an underground detector. This would be a detector that picks up Čerenkov radiation from muons as they travel through water, but frozen solid in the deep ice of Antarctica. The team has already performed tests in Arctic ice in Greenland, where in August 1990 they successfully detected muons with a short string of three phototubes suspended in a bore hole extending 100 metres below the packed snow layer, and about 200 metres below the surface.

An alternative way of using ice to detect ultra-high-energy neutrinos is being investigated by I.M. Zheleznykh from the Academy of Sciences in Moscow and others. They plan to 'listen' for radio waves produced by the Čerenkov effect, with the aim of detecting the whole burst of charged particles characteristic of the interaction of a high-energy neutrino. (The large number of particles involved would compensate for the small power radiated as radio waves.) In RAMAND, the Radio Antarctic Muon and Neutrino Detector, radio antennas on the surface would detect the radio signals produced by neutrino interactions in the ice below.

While Koshiba, Halzen, Zheleznykh and their colleagues dream of extending the present limits of neutrino astronomy here on Earth, others have let their dreams take them beyond the Earth itself. As long ago as 1965, Fred Reines for one saw the advantages in moving to the Moon to study cosmic neutrinos.

One of the major difficulties in studying cosmic neutrinos comes from the background of neutrinos that are produced by the interactions of charged cosmic-ray particles high in the atmosphere. In particular, charged pions and kaons created in these interactions emit neutrinos when they decay. At ultra-high energies, greater than 10^{12} eV, the problem becomes less severe because the effect of time dilation (see Chapter 4) makes the charged parents live longer and they can interact in the atmosphere before they decay. For this reason, detectors such as DUMAND will be useful only for neutrinos with energies greater than about 10^{12} eV. But at higher energies, the numbers of cosmic neutrinos are less, and these detectors must wait very patiently for rare events. It would clearly be to some advantage to remove the atmospheric neutrinos – in other words, to get rid of the atmosphere!

This is precisely what would happen with a neutrino experiment on the Moon. The experiment would still have to be under some lunar soil, or regolith, to provide a layer to absorb the primary cosmic-rays – rather as the atmosphere does on Earth. But in this case, the regolith would be dense enough to absorb secondary charged pions and kaons before they could decay, so the protective blanket would not itself become a source of neutrinos. Natural caverns, or 'lava tubes', 10 metres or so below the lunar surface, have already been considered as possible locations for a future manned base on the Moon, as the regolith would provide a natural way to shield personnel from cosmic radiation and meteorites. A large cavern of this sort might be suitable for a neutrino telescope.

Maurice Shapiro from the University of Maryland and Rein Silberberg from the Naval Research Laboratory in Washington, DC, in particular, have considered the advantages of lunar neutrino astronomy. They point out that it is in the detection of cosmic neutrinos with energies between 10^9 and 10^{12} eV that there is the most to gain. Although the probability for detecting a single neutrino in this energy band is less than it is for energies above 10^{12} eV, this is outweighed by the larger numbers of cosmic neutrinos at the lower energies. But on Earth, the energy band is made inaccessible by the background of neutrinos produced in the atmosphere.

Shapiro and Silberberg argue that a lunar observatory would open up the detection of neutrinos in the range of 10^9–10^{12} eV from broad, diffuse sources, such as the centre of our Galaxy, as well as from well-defined, discrete sources that may not emit enough neutrinos to be detected by DUMAND or similar observatories on Earth. Another possibility would be the detection of high-energy neutrinos produced in solar flares – eruptions at the Sun's surface that sends bursts of charged particles into space, which can provide a radiation hazard for astronauts. The detection of solar-flare neutrinos, especially *through* the Sun from flares on the far side, could provide a 'flare early warning system'.

The realisation of this dream undoubtedly lies well into the twenty-first century, as Michael Cherry from Louisiana State University has pointed out. The Moon's bulk could provide a 'target' for the cosmic neutrinos, but apparatus to detect the muons produced in the neutrino interactions would have to be built with materials brought from Earth. Even a detector of the most lightweight design conceivable today would require some 200 tonnes of material to be transported to the Moon.

Back firmly on Earth, other physicists have considered the role of neutrinos not in astrophysics, but in geophysics. The Earth itself is radioactive, a source of antineutrinos from the beta-decays of naturally occurring unstable nuclei. Each second, the lithosphere alone – the outer layer, some 1000 kilometres thick – is estimated to emit some 6 million antineutrinos through each square centimetre, mainly from the decays of uranium-238, thorium-232, rubidium-87 and potassium-40. This number of antineutrinos may seem large, but it is only a ten-thousandth the number of solar neutrinos that continually bombard the Earth. Experiments in neutrino geophysics would require apparatus ten thousand times as sensitive as the detectors designed to pick up solar neutrinos – an apparently impossible task, but one that does not prevent dreaming.

Suppose that you *could* build a detector sensitive enough to detect the Earth's antineutrinos? What could it tell you? Writing in *Nature* in 1984, theorists Lawrence Krauss, Sheldon Glashow and David Schramm considered the unique geophysical messages that the antineutrinos could convey. One advantage of these elusive particles is their great penetration, so that any measurements would reflect conditions within the Earth as a whole, not just in the surface layer, which is mainly all that is accessible in other ways.

The total number of antineutrinos emitted per second would reflect the overall radioactivity in the Earth, and therefore the amount of heat produced in this way. On the other hand, contributions from the different radioactive isotopes, which would be separated out by detecting antineutrinos in different energy bands, could provide windows on a variety of processes. Studying the amount of rubidium-87 in the whole Earth, for example, in contrast to the surface layers, could help in understanding convection currents in the mantle, 700 kilometres below the Earth's surface.

Another role for neutrinos in geophysics, considered first in 1974 by Georgii Zatsepin and L.V. Volkova, would be to use high-energy beams of neutrinos from accelerators to probe the varying density of the Earth, rather as X-rays reveal the varying density within a human body. The basic idea would be to direct neutrino beams across a whole range of angles through the Earth, and to detect the particles as they emerged on the opposite side.

However, writing in *Nature* in 1984, Thomas Wilson from NASA's Lyndon B. Johnson Space Center in Houston, argued that to work properly, as in a whole-body X-ray scanner, such a scheme requires that both the source of particles *and* the detector can be moved round the object. This is hardly practical with a large particle accelerator as a source and a detector like DUMAND! An alternative would be to have a network of detectors observing neutrinos from an astronomical source. The Earth's rotation would ensure that the source moved round the object, while the network of detectors would simulate the movement of a single piece of apparatus. Perhaps, if a strong enough high-energy source is ever found, this would be another application for Koshiba's world-wide network of 'lake' detectors!

From these few examples, it is clear that with neutrinos the future should be every way as exciting as the past. For more than 60 years, neutrinos have puzzled, intrigued, frustrated and surprised physicists whose interests range from the smallest structures inside the atom to the large-scale behaviour of the Universe. They may be something that is almost nothing, but they have already proved of immense importance in our understanding of many aspects of fundamental processes in the Universe. Neutrinos may not be absolute masters of the Universe – but they must surely rank as the most fascinating 'spaceships' in existence. I hope that you have enjoyed journeying with them.

FURTHER READING

This is a selection of books that I found interesting during my reading about neutrinos; some of them are no longer obtainable from bookshops, but should be available in libraries. The list of references for the quotations I have used in the book also contains many useful sources of information.

Books for the general reader

From Quarks to the Cosmos, Leon Lederman and David Schramm (Scientific American Books, New York, 1989). A particle physicist and a cosmologist join forces in an introduction to these two deeply interconnected areas of science.

The Particle Explosion, Frank Close, Michael Marten and Christine Sutton (Oxford University Press, Oxford, 1987). A highly illustrated introduction to the discoveries of particle physics, from the electron to the W and Z particles.

End in Fire, Paul Murdin (Cambridge University Press, Cambridge, 1990). A good account of the supernova SN1987A by a leading British astronomer.

The First Three Minutes, Steven Weinberg (Basic Books, New York, 1977). A relatively old but excellent account of the early stages of the Universe.

More technical books about neutrinos

The Neutrino, James S Allen (Princeton University Press, New Jersey, 1958). Possibly the first book devoted to this elusive particle, it contains interesting material, although some of it is out of date.

The Theory of Beta-decay, Charles Strachan (Pergamon Press, Oxford, 1969). A useful guide to Fermi's theory, and its subsequent development, with reproductions of papers on important theoretical and experimental advances.

Physics of Massive Neutrinos, Felix Boehm and Peter Vogel (Cambridge University

Press, Cambridge, 2nd edn, 1992). A good guide to modern experimental and theoretical work on neutrinos.

Neutrino Astrophysics, John Bahcall (Cambridge University Press, Cambridge, 1989). Mainly about solar neutrinos, by one of the leading experts in the field.

There are also at least two conferences a year dedicated to neutrinos; the proceedings of these and other conferences on particle physics are mines of up-to-the-minute information.

Neutrino Physics, ed. Klaus Winter (Cambridge University Press, Cambridge, 1991). A comprehensive overview of neutrino physics, with reprints of important early papers, and expert reviews of all areas of modern neutrino physics, with an emphasis on particle physics.

Books about the historical aspects of particle physics

Proceedings of the 'International Colloquium on the History of Particle Physics, July 1982, Paris', published in *Journal de Physique*, volume 43, supplement C8, 1982. Interesting talks by Fred Reines and Bruno Pontecorvo, among many others.

The Birth of Particle Physics, edited by L.M. Brown and L. Hoddeson (Cambridge University Press, Cambridge, 1983). Proceedings of a symposium on particle physics from 1930 to 1950, attended by many physicists who were active at that time.

Pions to Quarks, edited by L.M. Brown, M. Dresden and L. Hoddeson (Cambridge University Press, Cambridge, 1989). Proceedings of the 2nd International Symposium on the History of Particle Physics, held at Fermilab in May 1985. This follows on from the previous book, by covering particle physics in the 1950s, again mainly through the reminiscences of the people who were involved.

Inward Bound, Abraham Pais (Oxford University Press, Oxford, 1986). A physicist's account of the development of particle physics, especially up to the 1950s, written with great insight.

SOURCES OF QUOTATIONS

Chapter 1

1. H. Harari, *Proc. 13th Int. Conf. on Neutrino Physics and Astrophysics, Boston (Medford), June 5–11, 1988*, p. 574. © World Scientific Publishing Company.

Chapter 2

1. B. Pontecorvo, 'Proc. Int. Colloqium on the History of Particle Physics', *J. de Phys.* **43**, suppl. C8, 221, 1982.
2. W. Pauli in a letter to L. Meitner *et al*, reproduced in the original German in *Collected Scientific Papers by Wolfgang Pauli* (Wiley Interscience, New York, 1964), vol 2, ed. R. Kronig and V.F. Weisskopf, p. 1316. This translation is from *Inward Bound* (Oxford University Press, Oxford, 1986), by A. Pais, p. 315, and is reproduced here by permission of Oxford University Press.
3. W. Pauli, *ibid*.
4. E. Rutherford in a letter to O. Hahn, reproduced in *Rutherford* (Cambridge University Press, Cambridge, 1939), by A.S. Eve, p. 207.
5. E. Rutherford and H. Geiger, *Proc. Roy. Soc.* **A81**, 162, 1908.
6. L. Meitner, *Bull. At. Sci.* p. 2, November 1964. Reprinted by permission of the *Bulletin of the Atomic Scientists*, a magazine of science and world affairs; © 1964 by the Educational Foundation for Nuclear Science, 6042 South Kimbark Avenue, Chicago, IL 60637, USA.
7. L. Meitner, *ibid*.
8. J. Chadwick in a letter to E. Rutherford, Cambridge University Manuscript Collection, reproduced with permission of the Syndics of the University Library, Cambridge.
9. C.D. Ellis and W.A. Wooster, *Proc. Roy. Soc.* **A117**, 109, 1928.
10. B. Russell, *The ABC of Atoms* (Kegan Paul, Trench, Trubner & Co, London, 1923), p. 142.

11. N. Bohr in a letter to R.H. Fowler reproduced in *Niels Bohr – Collected Works* (North Holland, Amsterdam, 1986), vol 9, ed. R. Peirels, p. 555.
12. N. Bohr, *J. Chem. Soc.* **135**, 349, 1932.
13. N. Bohr, *ibid.*
14. N. Bohr, *ibid.*
15. N. Bohr, *ibid.*
16. N. Bohr, *ibid.*
17. W. Pauli, remarks at the Seventh Solvay Conference October 1933, reproduced in the original French in *Collected Scientific Papers by Wolfgang Pauli* (Wiley Interscience, New York, 1964), vol 2, ed. R. Kronig and V.F. Weisskopf, p. 1319.
18. N. Bohr in a letter to W. Pauli reproduced and translated in *Niels Bohr – Collected Works* (North Holland, Amsterdam, 1986), vol 6, ed. U. Hoyer, p. 443.
19. W. Pauli in a letter to N. Bohr reproduced and translated in *Niels Bohr – Collected Works* (North Holland, Amsterdam, 1986), vol 6, ed. U. Hoyer, p. 446.
20. W. Pauli, see Ref. 2.
21. J. Chadwick, *Proc. Roy. Soc.* **A136**, 692, 1932.
22. J. Chadwick, reprinted by permission from *Nature* **129**, 312, 1932, © 1932 Macmillan Magazines Ltd.
23. J. Chadwick, *ibid.*
24. D. Iwanenko, *Comptes Rendus* **195**, 439, 1932.
25. P.A.M. Dirac in *The Birth of Particle Physics* (Cambridge University Press, Cambridge, 1983), ed. L.M. Brown and L. Hoddeson, p. 49.
26. D. Iwanenko, see Ref. 24.
27. P.A.M. Dirac, see Ref. 25.
28. N. Bohr in a letter to F. Bloch reproduced and translated in *Niels Bohr – Collected Works* (North Holland, Amsterdam 1986), vol 9, ed. R. Peierls, p. 541.
29. D. Iwanenko, see Ref. 24.
30. F. Perrin, *Comptes Rendus* **197**, 625, 1933.
31. E. Fermi in *Z. für Physik* **88**, 161, 1934. This translation reprinted with permission from *The Theory of Beta Decay* by C. Strachan, p. 107, © 1969, Pergamon Press plc.
32. E. Fermi, *ibid.*
33. E. Fermi, *ibid.*

Chapter 3

1. H.R. Crane, *Rev. Mod. Phys.* **20**, 278, 1948.
2. R. Peierls, *Contemporary Physics* **24**, 221, 1983.
3. R. Peierls, *ibid.*
4. A.S. Eddington, *The Philosophy of Physical Science* (Cambridge University Press, Cambridge 1939), p. 112.
5. F. Reines, 'Proc. Int. Colloqium on the History of Particle Physics', *J. de Phys.* **43**, suppl. C8, 237, 1982.
6. F. Reines, *ibid.*
7. F. Reines, *ibid.*
8. F. Reines, *ibid.*
9. F. Reines, *ibid.*
10. F. Reines and C.L. Cowan, *Phys. Rev.* **92**, 830, 1953.
11. F. Reines, see Ref. 5.
12. F. Reines, *ibid.*
13. C.N. Yang, Nobel Lecture, 11 December, 1957. © The Nobel Foundation 1958.
14. C.N. Yang, *ibid.*
15. T.D. Lee and C.N. Yang, *Phys. Rev.* **104**, 254, 1956.
16. T.D. Lee and C.N. Yang, *ibid.*

17. T.D. Lee, unpublished.

18. T.D. Lee, unpublished.

19. W. Pauli, in a letter to V.F. Weisskopf, reproduced and translated in *Collected Scientific Papers by Wolfgang Pauli* (Wiley Interscience, New York, 1964), vol 1, ed. R. Kronig and V.F. Weisskopf, p. xiii.

20. A. Salam, *Imperial College Inaugural Lectures (1956–7, 1957–8)*, p. 54.

21. A. Salam, Nobel Lecture, 8 December 1979. © The Nobel Foundation 1980.

22. A. Salam, *ibid.*

23. A. Salam, *ibid.*

24. A. Salam, *ibid.*

25. M. Goldhaber, *1958 Ann. Int. Conf. on High Energy Physics at CERN* (CERN, Geneva, 1958), p. 233.

26. J.L. Vuilleumier, *Rep. Prog. Phys.* **49**, 1293, 1986.

27. W. Pauli in a letter to L. Meitner *et al*, reproduced in the original German in *Collected Scientific Papers by Wolfgang Pauli* (Wiley Interscience, New York, 1964), vol 2, ed. R. Kronig and V.F. Weisskopf, p. 1316. This translation is from *Inward Bound* (Oxford University Press, Oxford, 1986), by A. Pais, p. 315, and is reproduced here by permission of Oxford University Press.

28. E. Fermi in *Z. für Physik* **88**, 161, 1934. This translation reprinted with permission from *The Theory of Beta Decay* by C. Strachan, p. 107, © 1969, Pergamon Press plc.

29. J.L. Vuilleumier, see Ref. 27.

30. M. Fritschi *et al*, *Phys. Lett.* **B173**, 485, 1986.

31. M. Fritishi *et al*, *ibid.*

32. S. Boris *et al*, *Phys. Rev. Lett.* **58**, 2019, 1987.

33. J.F. Wilkerson *et al*, *Phys. Rev. Lett.* **58**, 2023, 1987.

34. B. Pontecorvo, 'Proc. Int. Colloquium on the History of Particle Physics', *J. de Phys.* **43**, suppl. C8, 221, 1982.

35. E. Fermi, quoted by B. Pontecorvo in Ref. 36.

36. E. Majorana, *Nuo. Cim.* **5**, 171, 1937, as translated by B. Pontecorvo in Ref. 36.

37. E. Majorana, *ibid.*

38. B. Pontecorvo, see Ref. 36.

39. B. Pontecorvo, *ibid.*

40. J. Bernstein, *Neutrino Cosmology*, CERN report 84-06 (CERN, Geneva, 1984), p. 7.

Chapter 4

1. C.D. Anderson in *The Birth of Particle Physics* (Cambridge University Press, Cambridge, 1983), ed. L.M. Brown and L. Hoddeson, p. 146.

2. H. Yukawa, *Tabibito* (World Scientific, Singapore, 1982), p. 190.

3. H. Yukawa, *ibid* pp. 194 and 195.

4. H. Yukawa, *ibid* p. 202.

5. H. Yukawa, *ibid* p. 203.

6. L. Lederman, *Scientific American*, March 1963, p. 60.

7. B. Pontecorvo, 'Proc. Int. Colloquium on the History of Particle Physics', *J. de Phys*, **43**, suppl. C8, 221, 1982.

8. B. Pontecorvo, *ibid.*

9. B. Pontecorvo, *ibid.*

10. F. Reines, 'Proc. Int. Colloquium on the History of Particle Physics', *J. de Phys.* **43**, suppl. C8, 237, 1982.

11. M. Schwartz, *Adventures in Experimental Physics*, (World Science Communications, New Jersey, 1972), vol α, ed B. Maglich, p. 82.

12. L.M. Lederman, *Proc. Int. Conf. on Instrumentation for High-Energy Physics* (University of California, Berkeley, 1961), p. 201.

13. L.M. Lederman, *Fermilab Report*, October 1988, p. 3.
14. G. Danby *et al*, *Phys. Rev. Lett.* **9**, 36, 1962.
15. M.L. Perl, *The Science Teacher*, December 1980, p. 16. © 1980, National Science Teachers Association.
16. M.L. Perl, *ibid*.
17. S. Weinberg, *Int. J. Mod. Phys. A* **2**, 301, 1987. © World Scientific Publishing Company.
18. J. Bernstein, *Neutrino Cosmology*, CERN report 84–06 (CERN, Geneva, 1984), p. 23.
19. M. Goldhaber, *Neutrinos – 1974* (American Institute of Physics, New York, 1974), AIP Conf. Proc. No. 22, p.1.
20. A. Pais in *Inward Bound* (Oxford University Press, Oxford, 1986), p. 521.
21. F. Halzen and K. Mursula, *Phys. Rev. Lett.* **51**, 857, 1983.

Chapter 5

1. L.M. Lederman, *High Energy Physics* (Academic Press, New York, 1967), vol II, ed. E.H.S. Burhop, p. 303.
2. C.A. Ramm, *CERN Courier* **6**, 211, 1966.
3. B.C. Barish, *Scientific American*, August 1973, p. 30.
4. A. Rousset, in a talk at the Centre d'Orsay de l'Université de Paris Sud, 12 March 1975, transcribed in *Hommage à André Lagarrigue*, p. 13.
5. P. Musset, in a talk at the Int. Colloq. on High Energy Neutrino Physics at the Ecole Polytechnique on 20 March 1975, transcribed in *Hommage à André Lagarrigue*, p. 35.
6. P. Musset, *ibid*.
7. *CERN Courier* **10**, 382, 1970.
8. D.H. Perkins, *Proc. 1973 School for Young High Energy Physicists*, Rutherford Laboratory report 74–38, p. vi–1. © SERC 1974.
9. S. Coleman, *Science* **206**, 1290, 1979. © 1979 by the AAAS.
10. S. Weinberg, Nobel Lecture, 8 December 1979. © The Nobel Foundation 1980.
11. S. Weinberg, *ibid*.
12. A. Salam, Nobel Lecture, 8 December 1979. © The Nobel Foundation 1980.
13. A. Rousset, *Neutrinos – 1974* (American Institute of Physics, New York, 1974), AIP Conf. Proc. No. 22, p. 141.
14. D.H. Perkins, *Proc. 1974 CERN School of Physics*, CERN report 74–22 (CERN, Geneva, 1974) p. 180.
15. R.P. Feynman, *Neutrinos – 1974* (American Institute of Physics, New York, 1974), AIP Conf. Proc. No. 22, p. 300.
16. W.K.H. Panofsky, *Proc. 14th Int. Conf. on High-Energy Physics* (CERN, Geneva, 1968), p. 23.
17. J. Bjorken, *Proc. Int. School of Physics 'Enrico Fermi', Course XLI* (Academic Press, New York and London, 1968), p. 55. © 1968 Societa Italiana di Fisica.
18. R.P. Feynman, *Science* **183**, 601, 1974. © 1974 by the AAAS.
19. R.P. Feynman, *Proc. 5th Hawaii Topical Conf. in Particle Physics*, p. 3. © 1974 by the University Press of Hawaii, Honolulu.
20. R.P. Feynman, see Ref. 18.
21. R.P. Feynman, *ibid*.
22. J. Steinberger, *Proc. 12th SLAC Summer Inst. on Particle Physics, June 1981: The Sixth Quark – Pief Fest*, SLAC report 281, p. 691.
23. J. Steinberger, *ibid*.
24. D.H. Perkins, *Introduction to High Energy Physics* (2nd edition), p. 275. © 1982 Addison-Wesley Publishing Co. Inc., Reading, Massachusetts. Reprinted with permission of the publisher.
25. P.V. Landshoff, *Proc. 1974 CERN School of Physics*, CERN report 74–22

(CERN, Geneva, 1974), p. 123.
26. R.P. Feynman, see Ref. 15.
27. D.H. Perkins, *Proc. 5th Hawaii Topical Conf. in Particle Physics*, p. 507. © 1974 by the University Press of Hawaii, Honolulu.
28. D.H. Perkins, see Ref. 14.
29. R.P. Feynman, see Ref. 18.
30. R.P. Feynman, see Ref. 15.
31. D.H. Perkins, see Ref. 24, p. 272.

Chapter 6

1. P. Morrison, *Scientific American*, August 1962, p. 91.
2. A.S. Eddington, *The Nature of the Physical World* (Cambridge University Press, Cambridge, 1929), p. 165.
3. H.A. Bethe, *The Sciences*, October 1980, p. 6.
4. J.N. Bahcall and R. Davis Jr, *Essays in Nuclear Astrophysics* (Cambridge University Press, Cambridge 1982), ed. C.A. Barnes, D.D. Clayton and D. Schramm, p. 243, also reproduced in *Neutrino Astrophysics* (Cambridge University Press, Cambridge, 1989) by J.N. Bahcall, p. 487.
5. F. Reines, reproduced, with permission, from the *Annual Review of Nuclear Science*, vol. 10, p. 1, © 1960 by Annual Reviews Inc.
6. J.N. Bahcall and R. Davis Jr, see Ref. 4.
7. J.N. Bahcall and R. Davis Jr, *ibid*.
8. J.N. Bahcall and R. Davis Jr, *ibid*.
9. J.N. Bahcall, *Scientific American*, July 1969, p. 29.
10. J.N. Bahcall and R. Davis Jr, see Ref. 4.
11. J.N. Bahcall and R. Davis Jr, *ibid*.
12. J.N. Bahcall and R. Davis Jr, *ibid*.
13. L. Wolfenstein and E.W. Beier, *Physics Today*, July 1989, p. 28.
14. K.S. Hirata *et al*, *Phys. Rev. Lett.* **65**, 1297, 1990.
15. J.N. Bahcall, reprinted by permission from *Nature* **330**, 318, 1987. © 1987 Macmillan Magazines Ltd.
16. P. Rosen, *'86 Massive Neutrinos: Sixth Moriond Workshop of XXIst Rencontres de Moriond* (Editions Frontieres, Gif-sur-Yvette, 1986), ed. J. Tran Thanh Van, p. 25.
17. J.N. Bahcall, *Neutrino Astrophysics* (Cambridge University Press, Cambridge, 1989), p. 31.
18. L. Wolfenstein and E.W. Beier, see Ref. 13.
19. J.N. Bahcall, see Ref. 17.
20. H.H. Chen, *Phys. Rev. Lett.* **55**, 1534, 1985.
21. S. Weinberg, *Int. J. Mod. Phys. A* **2**, 301, 1987. © World Scientific Publishing Company.
22. N.E. Booth *et al*, *Solar Neutrinos and Neutrino Astronomy* (American Institute of Physics, New York, 1984), AIP Conf. Proc. No. 126, p. 216.

Chapter 7

1. S.E. Woosley and M.M. Phillips, *Science* **240**, 750, 1988. © 1988 by the AAAS.
2. R. Jedrzejewski quoted by R.A. Schorn in *Sky & Telescope*, May 1987, p. 470.
3. R. Jedrzejewski, *ibid*.
4. S.E. Woosley and T. Weaver, *Scientific American*, August 1989, p. 24.
5. S.E. Woosley and T. Weaver, *ibid*.
6. A. Burrows, *Physics Today*, September 1987, p. 28.
7. S.E. Woosley and T. Weaver, see Ref. 4.

8. A. Burrows, *Proc. 13th Int. Conf. on Neutrino Physics and Astrophysics, Boston (Medford), June 5–11, 1988*, p. 142. © World Scientific Publishing Company.

9. R.M. Bionta *et al*, *Phys. Rev. Lett.* **58**, 1494, 1987.

10. K. Hirata *et al*, *Phys. Rev. Lett.* **58**, 1490, 1987.

11. S.E. Woosley and T. Weaver, see Ref. 4.

12. A. Burrows, see Ref. 8.

13. L.B. Okun, *Proc. 13th Int. Conf. on Neutrino Physics and Astrophysics, Boston (Medford), June 5–11, 1988*, p. 828. © World Scientific Publishing Company.

14. L.B. Okun, *ibid*.

15. L.B. Okun, *ibid*.

16. K. Greisen, *Proc. Int. Conf. on Instrumentation for High-Energy Physics* (University of California, Berkeley, 1961), p. 209.

17. F. Reines and J.P.F. Sellschop, *Scientific American*, February 1966, p. 40.

18. K. Greisen, see Ref. 16.

19. M.A. Markov, *Proc. 1960 Ann. Int. Conf on High Energy Physics at Rochester* (University of Rochester, New York, 1960), p. 578.

20. J.G. Learned, private communication.

21. K. Greisen, see Ref. 16.

22. J.G. Learned, *2nd Int. Workshop on Neutrino Telescopes*, Venezia, February 1990.

23. F. Reines and J.P.F. Sellschop, see Ref. 17.

24. F. Reines and L. Price, *Proc. Workshop on Particle Astrophysics: Forefront Experimental Issues*, p. 328. © World Scientific Publishing Company.

25. S. Weinberg, *The First Three Minutes*, p. 155. © 1977 by Steven Weinberg. Reprinted by permission of Basic Books, Inc., a division of HarperCollins Publishers.

26. G. Gamow, *The Creation of the Universe* (Viking Press, New York, 1952), p. 61.

27. A.S. Eddington, *The Expanding Universe* (Cambride University Press, Cambridge, 1933), p. 13. (Reprinted in 1987 in Cambridge Science Classics.)

28. R.A. Alpher and R.C. Herman, *Nature* **162**, 774, 1948.

29. S. Weinberg, see Ref. 25, p. 131.

30. S. Weinberg, see Ref. 25, p. 132.

31. D.N. Schramm and G. Steigman, *Scientific American*, June 1988, p. 44.

32. V.F. Shvartsman, *JETP Letters* **9**, 184, 1969.

33. G. Steigman, D.N. Schramm and J.E. Gunn, *Phys. Lett.* **66B**, 202, 1977.

34. J. Yang *et al*, *Astrophysical J.* **281**, 493, 1984.

35. S. Weinberg, see Ref. 25, p. 118.

36. G. Gelmini, *Neutrinos*, ed. H.V. Klapdor, p. 312. © Springer-Verlag, Berlin, Heidelberg 1988.

Chapter 8

1. M. Koshiba, *Physics Today*, December 1987, p. 42.

INDEX

240